The Semi-arid World

Geographies for Advanced Study
Edited by Professor Stanley H. Beaver, M.A., F.R.G.S.

1-10-75

The Semi-arid World:
man on the fringe of the desert

Dov Nir

The Hebrew University of Jerusalem

Translated from Hebrew by
Rivka Gottlieb

Longman

Longman
1724-1974

LONGMAN GROUP LIMITED
London
and LONGMAN INC., New York
Associated companies, branches and representatives throughout the world

English version © Longman Group Limited 1974

First published, in Hebrew, 1973 by the
Bialik Institute, Jerusalem

English version first published 1974

ISBN 0582 48067 1

Library of Congress Catalog Card Number: 73–89855

'Jadis on pouvait arguer du climat, de la végétation, du sol, de la topographie ou de la géologie locales pour expliquer les inégalités de richesse et de prospérité entre les diverses parties du monde habité. De nos jours ce genre de raisonnement est plus difficile à faire accepter. Les progrès de la science et des techniques les plus diverses donnent à un peuple, *qui veut et sait s'organiser,* une énorme capacité à modeler la nature, à réfondre les données physiques, à créer des ressources dans les cadres géographiques les plus variés.'

Jean Gottmann

Contents

List of Figures

List of Plates

Acknowledgements

Among the persons with whom the author has had the privilege of discussing the subject matter of this book Professor D. H. K. Amiran, who helped and took a lively interest in the work throughout its various stages, should be specially mentioned. The author's gratitude and deep respect for him are hereby expressed.

DOV NIR

Jerusalem
July 1972

Foreword

This book discusses man living in the semi-arid environment, and is essentially a study in regional geography. It is therefore necessary to define the terms 'regional geography' and 'the semi-arid zone'.

'Region' may be defined as a certain area which is the focus of a certain problem. The boundary between neighbouring regions is the area where one problem is becoming less acute, while another, which is dominant in an adjacent area, is beginning to show. The central problem in the region at hand is aridity, which means deficiency and unreliability of precipitation.

Research in regional geography regards the geographical conditions existing in the research areas as challenges for man; it examines man's activity in confrontation with these challenges, evaluates this activity and points out the various possibilities for solving the central problem of the region. The semi-arid zone well illustrates the scope of regional geography. It contains a wide variety of relationships and responses to challenges ranging from absolute fatalism, through sincere but technologically poor efforts, to the most sophisticated enterprises of our time: why is it that man reacts so differently to the same challenge in regions with almost identical natural conditions? The purpose of this book is to attempt to answer this question: to review the various possibilities of which only one or the other has been chosen, since the importance attached by man to the organising of his environment varies.

It seems to the author that a study of this kind may contribute towards comprehending the complex processes at work in the relation between man and his environment, man's reactions to the challenges of his environment and his rising above the basic conditions which restrict him. The three main variants of the arid zones—the arid, the semi-arid and extreme arid[1]—comprise 36 per cent of the total area of the continents, but the population living in this area in 1960 was no more than 13 per cent of that of the world (Amiran, 1966). Nevertheless, these 13 per cent are 380 million—and the destiny of 380 million people deserves the attention of the student of geography.

[1] A discussion of definitions and coefficients for arid zone classification is included in the Appendix (p. 167).

Introduction: The ecumenic importance of the semi-arid zones and their problems

The semi-arid zone occupies 43·5 per cent of the entire area included within the boundaries of aridity, that is, the area where absence or scarcity of rain is the first natural obstacle and where the main challenge to man is the effort required to supply water. The arid zone occupies 44·6 per cent and the extreme arid zone 12 per cent of the entire arid area (Amiran, 1966), but 72 per cent of the population live in the semi-arid zone, whereas only a little more than 1 per cent live in the extreme arid zone. These figures indicate the importance of the semi-arid zone, in which 276 million people live (1960), comprising 8 per cent of the total world population, with a population density of 13 people per square kilometre (34 per sq mile). The arid and extreme arid zones, though they are important arenas of special geomorphological processes, have but limited ecumenic importance, and will not be dealt with in the present work.

Figure 0.1 illustrates population distribution in the various arid zones. The semi-arid fringes are relatively densely populated, whereas the arid interior, and particularly the extreme arid centre, are sparsely populated, except in isolated cases where there is an aiding factor, such as a great river, which encourages a dense population.

Is the semi-arid zone population static and backward, or is it dynamic and active? The density values in themselves are not a coefficient of a certain area's dynamic. We consider rate of density growth during a certain period as the coefficient which indicates a population's dynamic. It appears that population growth rate varies in various regions in the semi-arid zone; aridity should not, therefore, be taken as the only factor determining accelerated or moderate increase in density.

During the years 1920 to 1960 population density in Sicily grew by 24 per cent and by the end of that period reached the ratio of 183 persons per sq km (476 per sq mile); in the Alicante district of Spain density increased by 38 per cent (121 persons per sq km: 315 per sq mile) and in the Seville district by 95 per cent (88 persons per sq km: 229 per sq mile); in Argentina

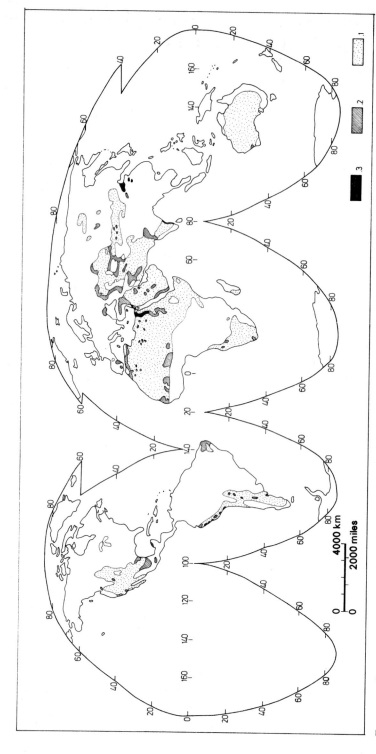

Fig. 0.1. Population distribution in arid zones (Hills, 1966). 1. Less than 10 per sq km (25 sq mile) 2. 10–100 per sq km (25–250 sq mile). 3. Over 100 per sq km (250 sq mile).

by 172 per cent (35 persons per sq km: 91 per sq mile); and in California by 454 per cent (38·6 persons per sq km: 100 per sq mile) (Hodge, 1963). Although the increase in population density is inversely proportional to density values, it is most impressive in the three latter instances and testifies to the dynamism of these regions, as opposed to the backwardness of the first two. Since climatic conditions in the first sample regions are fairly equal—or at least, they have been classified as equal according to the accepted aridity coefficients—it appears that the reason for different growth rates is to be found in other factors, ethnic, political and economic. The explicit purpose of this work is to look for the answer to this question, to study and analyse the various measures which man has employed in order to overcome the climatic obstacle. The second element which makes the semi-arid zone an important ecumenic area is the fact of its being transitional to more humid regions with a relatively dense population which presses on its fringes. The semi-arid zone belongs, therefore, economically, and in many cases also politically, to countries with a better rainfall regime. Its marginal situation is advantageous for the exchange of products, availability of agricultural or industrial population, relatively short distances to marketing centres, and the possibility of dependence on a better developed economy.

The ethnic and economic structures of the semi-arid countries are diverse (Fig. 0.2).

The greater part of the arid zone is occupied by traditional economic frameworks based on hunting and gathering, or on dry farming. Even in the more developed political frameworks the economy is still based mainly on livestock and dry farming, though their technological levels vary. The transition towards a sophisticated agriculture is only in its initial stages, and there are very few industrial nuclei in this zone, except for two or three limited areas in no more than half a dozen countries.

Precipitation deficiency is by no means the only problem of the semi-arid countries. The solution of the water problem does not guarantee the solution of the economic and ethnic problems of the region; moreover, it will bring other factors to the surface: the level of education and technology, that can facilitate (or hinder) the execution of proposed plans; the standard of living, purchasing power, the standard of marketing, the habits, tradition, conservatism of the people—all these may, according to their respective natures, hasten or delay the solution of human problems.

Our discussion cannot, therefore, be a technological study of irrigation; it has to be a thorough geographical review, an attempt to distinguish those factors which are dominant, and to find the links between them, even though they cannot always be measured quantitatively and expressed in mathematical correlations. The delivery of water to the region does not necessarily guarantee its development, and does not in itself ensure a proper standard of water utilisation.

The semi-arid zones—or 'fringes of the desert' as they are here called—

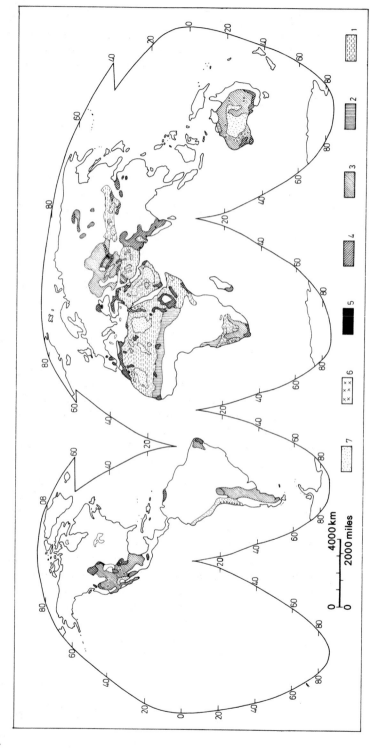

FIG. 0.2. Economic activity in arid zones (Hills, 1966). 1. Nomadic grazing. 2. Traditional agriculture, hunting. 3. Cattle farms. 4. Intensive agriculture and grazing. 5. Manufacture and industry. 6. Mining. 7. Low economic activity.

are among the most unstable areas with regard to population. Occasionally whole populations migrate as a result of catastrophes, such as a prolonged drought, or destructive floods, which recur in the arid zone's history, whether in central Asia or in the Mediterranean 'fertile crescent' in past times, or in the twentieth century in the 'dust bowl' in the mid-west of the United States. Because of the delicacy of water installations—irrigation ditches, wells, water holes—the absence or existence of an organising authority is particularly felt here; this is the most sensitive region with regard to hostile invasions and their destructive consequences.

Water is the major factor of subsistence in this zone, perhaps even more so than soil, especially where technology is undeveloped. But when the level of technology and the standard of living rise, and it becomes necessary to maintain a modern economy which can compete with other regions, the problem of the relativity of the water's value arises. In an undeveloped state of technology the existence of man is based primarily on the securing of food supply; thus the side of the region which faces the desert will seem to the desert dwellers as 'a land flowing with milk and honey', whereas the side facing the humid region may seem to its dwellers as a backward, problematic, marginal and supported region. The future of the semi-arid region depends, therefore, on its capability of competing with neighbouring developed regions, and on its ability to make the most profitable use of water, which in the case of developed technology may not necessarily be a solution for food supply. The natural resources of the semi-arid zone—water, many hours of solar radiation, mild climatic conditions as compared to most humid zones—should be properly evaluated by a new global concept, which takes into consideration modern notions of space and economy. A thorough revaluation of the region seems to be necessary, particularly as regards the purposes for which water should be used, so that it can serve an area where the ecumene is expanding rather than shrinking.

This subject of aridity or semi-aridity concerns sixty countries (Batisse, 1969), that is, to about half the countries in the world. The sociological, economic and ethnic conditions are different in each of these countries and the answers to their problems will vary accordingly.

The natural common denominator of the arid zones is not exclusively negative; on the contrary, we have already pointed out several important factors which can increase their power of attraction. The various challenges exist; the problem is how to reconcile and organise the abundance of solar radiation and vacant space with water deficiency, and turn it to man's benefit.

PART ONE

The natural challenges which man has to face

1
Climatic problems

Irregularity in occurrence and intensity of rainfall

It is true that definitions characterising the arid zone are based on the general water balance; but they fail to define one of the essential features of the region; the uncertainty as to when rain will fall (Fig. 1.1). The arid zone is characterised not only by meagreness of precipitation (in annual average), but also by uncertainty as to when and in what amounts rain will fall. The average data that serve as a basis for the various coefficients used to define the arid zone's boundary, are an abstraction over many

Plate 1. Beer Sheva, a modern town on the fringe of the Negev Desert (courtesy: the Director, Department of Geography, Hebrew University of Jerusalem).

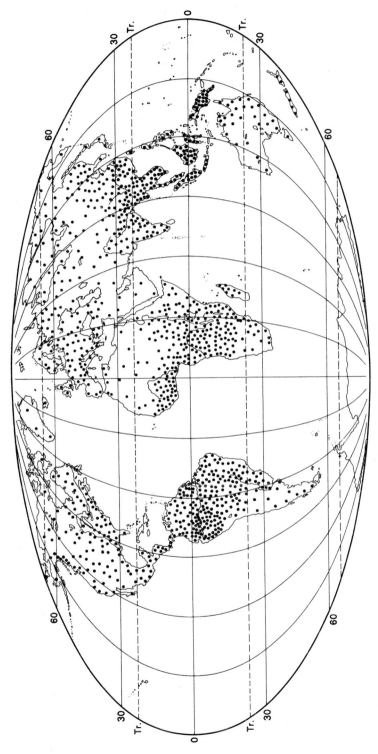

Fig. 1.1. World precipitation. Each point represents 100 km³ of precipitated water in one year (Péguy, 1961).

9

years, and have no meaning in relation to actual rainfall; but it is the actual amount of rainfall at a given time that determines the nature of a region, whether it is arid or not. A prolonged cessation of rain during the rainy sesason will result in the withering of young plants; it is the expression of the irregularity of precipitation. This irregularity, that is, the deviation from the annual average, increases in direct proportion to aridity. While the average annual deviation in the temperate or the humid equatorial zones does not exceed 10–20 per cent, it reaches 30–40 per cent in the arid zone, and the actual deviation (not the average) which is the one that affects the plants and the farmer—may even reach values of hundreds per cent.

A world map of precipitation variability (Fig. 1.2) illustrates clearly this characteristic of the arid zone. The annual variation is sometimes so great, that it is devoid of any statistical meaning; for instance, two extreme cases were recorded in Swakopmund, South-west Africa, in the course of twenty-five years—one of 1 mm and the other of 149 mm (5·9 in) in a year; there are, however, less extreme examples, which nevertheless still prove the undependability of rainfall, as illustrated below.

The Beth She'an area, in the Jordan valley, is by definition a transitional area between the Mediterranean climate (CSa) and the semi-arid, steppe climate (BS). The following table illustrates the annual variability at two stations, about 10 km (6 miles) apart (Nir, 1968):

Annual variability of precipitation in Kefar-Ruppin and Beth She'an in the years 1939–47

years	Beth She'an station annual amount mm	in	deviation from average in %	Kefar Ruppin station annual amount mm	in	deviation from average in %
1939–40	302	12	− 6	255	10	− 13
1940–41	352	14	+10	253	10	− 13
1941–42	259	10	−19	226	9	− 22
1942–43	514	20	+67	495	19·5	+ 68
1943–44	291	10·5	− 8	324	13	+ 11
1944–45	528	20·5	+68	652	25·5	+121
1945–46	253	10	−21	194	8	− 33
1946–47	155	6	−49	147	5·5	− 50
Average deviation			30			41

The average deviation for these years was 30 per cent in Beth She'an, which still belongs to the marginal Mediterranean climate, whereas in Kefar Ruppin, which is on the arid side of the aridity boundary, it reached 41 per cent. Actual deviations show that while in Beth She'an extreme values did not exceed 68 per cent deviation from the average, in Kefar

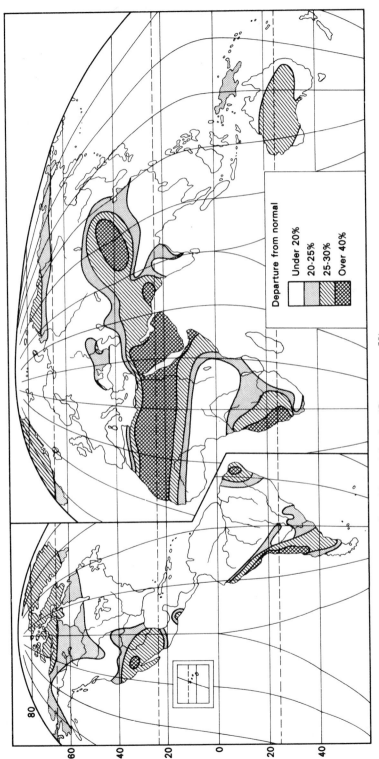

Fig. 1.2. Deviation from annual average of world precipitation (Rumney, 1968).

Departure from normal

Under 20%
20-25%
25-30%
Over 40%

Ruppin the deviations from the average reached 121 per cent.

The lower the annual average, the more critical for crops is the negative deviation from this average; and since, in addition, the deviation increases in direct proportion to increasing aridity, it becomes clear that the uncertain character of rain is a fundamental factor for crops. If, for instance, the average value at a certain station is 250 mm (9·8 in), the annual amount to be expected is (in the case of 20 per cent average deviation) 200–300 mm (7·8–11·8 in). In this case, the region is still capable of bearing crops (if, indeed, the deviation remains within the limits mentioned above). But if the annual average is only 150 mm (6 in), and the average deviation reaches 40 per cent, it is to be expected that actual amounts will fall short of the requirements of the necessary water balance of 90–250 mm (3·5–9·8 in). On the other hand, great amounts of rain are not always effective in a region that cannot absorb them. In western United States we find several indicative examples: a station in the Mojave desert, with an annual average of 55 mm (2 in) a year, received during three successive years only 0·25 mm, and this in a single shower, in March 1919. In 1921, 247 mm (8·8 in) of rain fell. Another case: Yuma (Arizona), with an average precipitation of 85 mm (3·5 in), received 235 mm (9·5 in) in 1905 and only 7·5 mm (0·3 in) in 1953 (Jaeger, 1957). All these occurred in the winter rains region; the summer rains region receives its precipitation in the ineffective form of thunderstorms: twelve separate local storms which occurred over an area of 300 sq km (116 sq miles) in the south-western United States showered rain over an area of only 7 sq km (2·7 sq miles).

Extreme cases occur in the northern parts of the Peruvian coast in South America. A branch of the equatorial countercurrent carries towards the northern coast exceedingly warm water, reaching sometimes as far as Lima and even Pisco. Every few years this current increases in magnitude, and extending southward to the coasts that are cooled by the Humboldt current, its ability to warm the air and create thermic instability becomes more pronounced. It results in unusually powerful showers over the arid northern coast of Peru. In Trujillo, between 1918 and 1925, the total amount of rain was 37 mm (1·5 in); but in March 1925 rain amounts reached 350 mm (14 in), 222 mm (8·7 in) of which fell in three days (Rumney, 1968). It is evident, therefore, that in such extreme cases it is meaningless to talk about a 'normal' amount.

In such unstable and uncertain rainfall regimes, the amount of rain should be evaluated differently: the irregularity of rainfall should be considered as well as its amount. Even if in a certain year the annual amount of rainfall does not greatly depart from the average, its value to agriculture may still be small because of irregular distribution; for instance, a concentration of intense showers at the beginning of the rainy season, followed by a pro-longed drought, will result in the withering of new plants; or, in case of a delay in rains at the beginning of the season, seeds will not germinate, and heavy rainfall at the end of the season will be too late to revive them. The length of the rainy season is important for the groundwater balance as well, at least for the horizon which is nearer to the surface. A distinction should, then, be made between 'meteorological' drought and 'effective' or 'agricultural' drought.

The definition of 'meteorological' aridity is the duration between two significant falls of rain. The duration of such periods has been known to reach several years in extreme desert regions; in the Sahara, as much as eight years, in Peru—eighteen years. In the semi-arid zone drought durations are shorter—five to eight months between two rain seasons; but there are also dry spells in the rainy season, stretching over days or even several weeks. As long as monthly deviations are smaller than the mean annual deviation, that is, the regional water balance is not distrubed, meteorological drought is meaningless; if, on the other hand, the deviations exceed the normal, it is regarded as a meteorological drought.

The effective aridity, in other words the drought, is a pragmatic concept, alluding to the duration between two rains that can feed groundwater. Such rains create surface runoff. The duration of drought is long in the extreme arid zone, and decreases towards the margins. Similarly, the difference between meteorological and effective drought becomes smaller. Still, even in a typical Mediterranean region, such as the northeast of Spain, or the various Mediterranean islands, the drought period may endanger the replenishing of the upper horizon of groundwater.

The great intensities of rain in the semi-arid zone, per minute as well as per hour and per day, are to be explained by the irregular nature of precipitation; intensities of 40–50 mm (1·5–2 in) per hour are not rare in the semi-arid zone; daily intensities of 60–80 mm (2·5–3 in) are also not unheard of. There is a certain degree of regularity in rain intensities in the Mediterranean climate: the greatest intensities occur at the beginning or end of the season, when warm and unstable air masses may invade the semi-arid zone; at the height of the Mediterranean winter, however, intensities are generally smaller, though the total amount of rain is greater (Nir, 1970).

Rainfall intensities in the coastal deserts are greater than inland intensities, as a result of the contact between humid maritime air masses and dry continental air masses; the same applies to the southern margins of the arid zone, where humid equatorial air masses come into contact with continental desert air masses.

Thermal aspects of the arid zone

In the prevailing conditions of cloudlessness, low relative humidity and absence of water vapour in the air, the solar radiation is very intense (Planhol and Rognon, 1970); the number of hours of sunshine is very near the theoretical maximum: in the Middle East, for instance, it reaches

80 per cent of the possible 4 500 hours. The following data provides some comparison with other climatic regions (Ashbel, 1950):

Manchester	900 hours
Munich	1 850 hours
Beirut	3 000 hours
Amman	3 650 hours

These values have a bearing on the flora and the fauna, including man, and on man's utilization of the many hours of insolation. Today, when a considerable portion of man's time is consumed in leisure and recreation, this feature of the arid zone is a source of attraction for holiday makers, and is acknowledged as one of its natural resources. Because of the intense insolation the highest temperatures in the world have been recorded not in the equatorial zone, where evaporation and cloudiness are great, but in regions nearer the Tropic of Cancer and the Tropic of Capricorn. In addition to intensity of radiation and high temperature values the long period through which these high values prevail is also of importance.

Because of the anticyclonic cells which carry away the atmospheric water vapour, the region is cloudless and therefore the air layer which is nearer to the ground becomes exceedingly hot, thus further preventing saturation of air with water vapour. The excessive evaporation creates an increased upward capillary movement of water in the soil. Since most of the water in the upper horizon is replenished only by rainwater, it follows that in the interval between two falls the soil is prone to dry up completely, and the more so between two rainy seasons. The depth of dry soil may reach at least one metre (3 ft), which means an immediate effect on plants with roots shorter than this.

Potential evaporation may reach in extreme cases several metres; even on high mountain peaks, reaching 3 000 m (9 840 ft) in the central Sahara, the evaporation may reach 5 m (16 ft). The relief has no influence in this case; in fact, the potential evaporation depends on the values of air saturation deficit near the ground. Even though the *absolute humidity* of the air near the ground exceeds the values characteristic of the temperate zone, the relative humidity, because of the high temperature, is quite small and very far from saturation point. Relative humidity values are usually less than 40 per cent, but in extreme cases—such as the spring hot spells (Sirocco, Khamsin, Sharav)—humidity falls to minimum values of 2 to 4 per cent. The greater the deficit of the air water content, the quicker the absorption of additional moisture brought by rain.

Additional climatic elements

This discussion may create a false notion that all the definitions of climate are negative, that is, with regard to human life there are here only negative values—absence of water and precipitation, lack of moisture etc. Yet, there are also climatic elements that may not be negative. Thus, for instance, the long hours of sunshine during most of the year may be turned into a source of attraction for tourism and recreation for the population of adjacent humid and cloudy regions, who can travel easily in the present economic and technological situation. The vacant unpopulated spaces constitute choice areas for experiments that require wide spaces; the research on radiation and the observation of artificial space objects are particularly convenient in the conditions of clear skies that prevail in the arid zone. The preservation of organic matter on the surface, as well as the durability of soluble minerals, in the arid climate conditions, enable their exploitation and utilization; similarly, the ability to practice open-mining of various resources is a great advantage of this climatic region. All these must be regarded as positive elements of the arid climate.

Pleistocene climates and their significance for the present landscape

Possible changes in wind systems

In order to understand the possible shift of the global wind systems which occurred during the Pleistocene, it is necessary to look, first, into the reasons for climatic aridity. Two factors are chiefly responsible for climatic aridity: relief and the earth's division into continents and oceans on the one hand, and the physical properties of the global wind system on the other (Butzer, 1961). The negative influence of the relief may be defined as the 'rain-shadow' created by a high mountain barrier which lies perpendicularly to the main air flow carrying humid air masses; this influence causes the rainshadow deserts—Takla Makan in central Asia, the plateau of the central Rocky Mountains, or the Judaea Desert in Israel. It is clear that if there was any significant shift of air flow during the Pleistocene, it could have had a considerable bearing on the existence of this kind of desert in those places. As for distance from the sea, the core of each continent, far removed from the warm and humid oceans, is dry in most cases. In central Asia, even if we ignore its mountainous frame, the moist air masses have to pass thousands of kilometres from their source in order to reach the core; by that time they have usually lost the moisture with which they started. These factors of relief and distance from the sea are responsible for aridity in the middle and high latitudes: the great plains, and the range and basin area in the western United States; Patagonia desert; the Eurasian desert extending from southern Russia to Mongolia; the interior of Iran. There is no reason to believe that essential climatic changes have occurred in these areas, since fundamental geographic phenomena, and not the wind system, are the basis of climatic conditions.

As for the world wind system, it is affected by two main factors: (1) the deflective force of the earth's rotation and the friction with the atmosphere; (2) the meridional north–south tendency of the heat balancing direction caused by the accumulation of solar insolation at the equator, and the loss of heat as a result of outgoing earth radiation, in the polar regions. Were it not for the earth's rotation the hot air would have ascended in the atmosphere and moved poleward, while cold air from the poles would have moved equatorward as a surface current. The earth's rotation gives the upper current a westerly component; but because of vortical changes in the southward bound air masses, the westerly flow increases and reaches its peak at the latitude of 40°; south of that latitude is the steepest thermic gradient between the equator and the poles, and also the maximum velocities at the upper layer of the atmosphere, known as the 'jet stream'; the force of the jet stream creates in the lower layers an easterly air flow. Westerly winds are thus dominant in the lower atmosphere between the latitudes 35° to 75°, while between 35° and the equator the prevalent direction of winds is easterly. In the zone of contact between the westerlies and the easterlies the air subsides and creates the subtropic belt of high pressure cells, in fact an assemblage of anticyclonal cells; between them and the equatorial low pressure zones there is a thermic exchange which forms the trade winds that flow over the oceans and the western parts of the continents. Within this belt cold and humid air masses from high latitudes come into contact with hotter and drier masses from lower latitudes; between these two kinds of air masses there is a decrease of temperature from 5° to 3°, and moisture may decrease from 50 to 30 per cent. This contact area, which is called 'the trade wind inversion', exists during the summer at an altitude of 700–800 m (2 300–2 625 ft) above the Mediterranean and at an altitude of 1 200 m–1 400 m (3 950–4 600 ft) above the North Atlantic Ocean. This contact area is the upper limit of convection clouds and is responsible for the meagreness of precipitation in the trade wind current. The deserts of Sahara and the Arabian Peninsula, the Thar desert in India–Pakistan, Namib and Kalahari in South Africa, the Atacama desert and the deserts of Peru and Bolivia, and the Australian desert—all these are within the boundaries of the trade wind regime, and their aridity is a result of conditions created by the global wind system. Only the monsoon of southeast Asia is powerful enough to shift the currents southward or northward, so that this region gets large amounts of rain, in spite of its geographical latitude.

Presentday fluctuations of the wind systems

The global wind system represents a balance between several forces which are an expression of the earth's rotation and solar radiation. While there is no significant change in earth's rotation from the climatic point of view, the solar insolation intensity undergoes periodic, and perhaps even episodic, changes. The most outstanding change is found in the difference between winter and summer, with reversed conditions of the global wind system, including the monsoons: this is the result of changes in solar insolation and in the thermal gradient in relation to the geographical location. The changes on the sun's surface, involving an increase or decrease in radiation, affect also the distribution of pressure centres and the global wind system; that is, slight shifts in the global wind system are possible and to be expected as an ordinary phenomenon, mainly as a result of activities on the sun's surface. In the long run such variability can achieve great dimensions. In other words, there is nothing surprising in climatic changes which occurred in the course of geological periods, if we are aware of contemporary fluctuations of short duration and extent. The short-term fluctuations of the global wind system which may last for days or even for several weeks, are the key for understanding the supposed fluctuations of the Pleistocene. Willet (1949) and Flohn (1952) divided these supposed fluctuations in two generalised patterns—the zonal and meridional. The *zonal pattern* is characterised by a well-developed jet stream, undisturbed by cold or warm air masses originating in higher or lower latitudes. The subtropic cell of high pressure and the subpolar centre of low pressure are well developed and the thermal gradient, that is, the jet stream retreats poleward. The flow of the trade winds is regular. To sum up, here the main flow of air has a west–easterly direction in the shape of the well-developed jet stream. The periods of the *meridional flow pattern*, on the other hand, are marked by a weakening of the global westerlies, while at the higher air layers there exist troughs, through which polar air masses invade far into the tropics, while, in exchange, tropic air masses are transferred far into the higher latitudes. In the low layers of the atmosphere the low pressure belt almost vanishes while the jet stream shifts to the south. The subtropic highs weaken and are separated by highly dynamic lows which interrupt the regular flow of the trade winds, thus permitting an invasion of cold air into the tropics and an increased activity in the intertropic front.

These two possible wind systems have a bearing on the Pleistocene climates in that each, when prevailing during a longer period, is related to one particular type of climate, either the glacial or the interglacial. The *zonal flow pattern* is related to the interglacial period: the tendency of the jet stream to shift northwards, with very little or complete absence of cold front intrusion into the tropics, does not encourage rain over the desert belt of the trade winds. The glacial periods, on the other hand (and the short duration weather changes for the worse during the seventeenth and nineteenth centuries resulting from the advance of glaciers), are regarded by Willet and Flohn as *meridional anomalies*. The disintegration of the anticyclonic cells, combined with an invasion of cold air masses into the tropics, stimulates rains over the trade wind belt, especially over the southern and northern margins. These two patterns existed, of course, during a definite period, since a full and continuous balance of heat transference from the equator to the poles cannot be maintained without fluctuation of this kind; but there are annual changes in the relative weight of each pattern.

The Pleistocene climates: pluvial and interpluvial periods

The Pleistocene includes five periods in which the ice sheets in higher latitudes grew and advanced, interrupted by four or five interglacial periods;
all these occurring during the last $1-1\frac{1}{2}$ million years. At the time when man
had not yet fixed his secure domain of subsistence, about 10 000 to 12 000
years ago, the fluctuations of glacial advance and retreat generated southward and northward migrations of animals and hunters. Today, too,
there are population migrations, especially where technological conditions
are inferior. Whole populations have been known to migrate from one
region to another as a result of droughts. The term 'pluvial' which indicates
an invasion of more humid conditions to the lower latitudes, was suggested
by Hull in 1884. The pluvial periods were parallel in time to the glacial
periods of the higher latitudes. The pluvial period is defined as a stage of
prolonged and widespread increase of precipitation, which is long and
intensive enough to be geologically significant. Ecologically, the Pleistocene
climates varied from one region to another (Fig. 1.3).

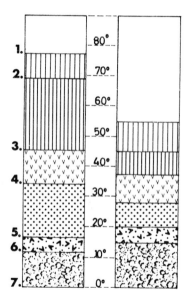

FIG. 1.3. Shifting of climatic zones during the Würm glaciation (after Büdel, 1957).
1. Glacier zone. 2. Tundra. 3. Non-tropical forest. 4. Mediterranean forest and steppe.
5. Desert. 6. Savanna. 7. Tropical forest.
On the left side: Holocene. On the right side: Würm.

In the arid zone of the middle latitudes the ice age has left its impression
on the Eurasian steppe, the Spanish Meseta, the Atlas range, etc. The region
is partly a planetary desert, partly a relief desert. The glacial climate
brought reduced temperatures, but not increased precipitation; still, the
lower summer temperatures reduced the evaporation, and in consequence

the level of the inland lakes was higher than it is today; the Caspian Sea, for instance, drained into the Black Sea. The vegetation, too, was of the temperate, and even subpolar type. River terraces were created of periglacial or fluvioglacial material, and cryoturbation processes had great influence upon the formation of soils; during the cold dry periods considerable layers of loess were deposited (as in southern Ukraine).

In the subtropic arid zone, including the Thar desert, the Near Eastern 'Fertile Crescent' and North Africa, the dominant factor was not diminished temperatures (by 4°C), but the increase in precipitation and prolongation of the rainy season; judging by the present size of the Dead Sea as compared to the maximum dimensions of the 'Lisan Lake' during the Würm period, precipitation seems to have increased by 30 to 40 per cent (Ben-Arieh, 1965). It follows, therefore, that during the pluvial periods man's dispersion could extend over wider areas, including regions that are now acknowledged deserts; even in the Sahara it was possible for animals and vegetation to exist (Fig. 1.4).

The pluvial periods in the tropical and equatorial arid zone have left land forms and fossil soils in an area which today is an absolute desert. On the other hand, the interpluvial periods left arid landforms, such as the seif dunes, in an area which today is typical steppe (Grove and Warren, 1968). In this region—southern Sahara, Rhodesia and parts of South Africa which also come under the influence of the anticyclonal cells—the significant difference is not in temperatures but in the amounts of precipitation which were much larger than they are today. The Sahel lakes' high water levels are classical examples; many fresh water lakes have survived between Tibesti and Ennedi; the level of Lake Chad was higher by 60 m than its present surface; this is proved by the existence of lacustrine soils, and even red tropical soils, in areas which today are typical deserts. Evidences of pluvial stages have been found also in the valleys of the Vaal and the Zambezi. Eyre lake, in Australia, which nowadays may dry up sometimes for several months, or even several years, extended over an area of more than 100 000 sq km (38 610 sq miles), and was more than 50 m (196 ft) deep.

The Holocene climate

The Holocene included the period that followed the last glacial and pluvial periods, starting about 12 000 years ago; the beginning of that period marked the formation of the arid conditions which have existed with fluctuations to this very day, and which characterise the semi-arid zone and constitute its common denominator that provides the justification for the present discussion. Still, various Mesolithic, Neolithic and historical findings testify clearly to favourable or unfavourable changes in the zonal conditions of aridity. Judging by geological and faunistic findings in the Sahara and in Israel, it seems that between 5000 and 2350 BC the region

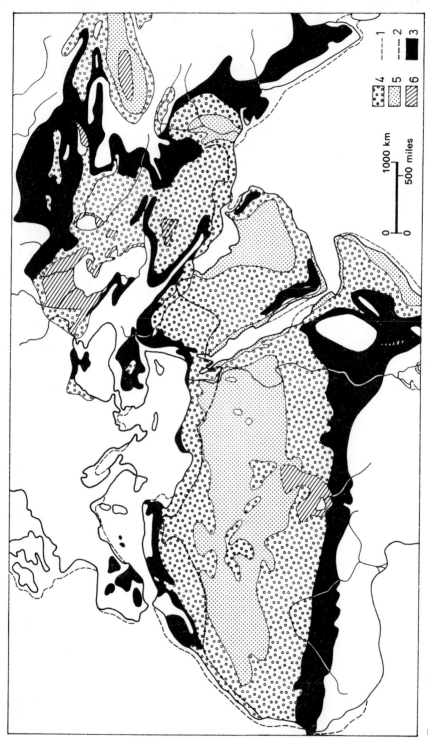

Fig. 1.4. Reconstruction of climatic conditions during the pluvial stage in the Near East and North Africa (Butzer, 1961). 1. Approximate shorelines (−100 m isobath). 2. 18°C isotherm during a pluvial phase. 3. Modern semi-arid zone belonging to humid climatic province in pluvial times. 4. Semi-arid zone in pluvial times. 5. Arid zone during pluvial times. 6. Extension of inland seas in pluvial times.

experienced more favourable conditions of precipitation. Stone drawings done by an ethnic group of hunters and cattle breeders depict not only their cattle, but also tropical fauna that is non-existent today in northern Sahara: elephants, giraffes, hippopotami, buffaloes, etc; for the existence of elephants and buffaloes a rainfall of not less than 150 mm (6 in) is necessary. Butzer assumes that the southern boundary of the Sahara has shifted a few hundred kilometres northward. Floristic tests, particularly pollen tests, have disclosed a Mediterranean vegetation on the high mountains in central Sahara (Quèzal, 1963). Similar evidences of this slight increase in precipitation exist also in New Zealand, in the Jordan Valley, and in the Carmel coastal plain in Israel. Those favourable conditions were again replaced by more arid conditions at the end of the third millennium BC.

The importance of this discussion dealing with climatic conditions in the arid zone, which were different from those existing today, is in the recognition that many of the phenomena existing in the region today—soils, landforms, river valleys, dry lake beds, various kinds of vegetation—are a heritage from different climatic conditions. Thus, when we come to discuss a region, we have also to examine its Pleistocene past in order to understand present circumstances. Much of the flora and fauna existing today in the semi-arid zone is residual from pluvial fauna and flora. Similarly, a certain part of groundwater resources was accumulated during the Pleistocene when precipitation was more abundant, and it is not replenished today. A large proportion of the soil—if not indeed all of it—was formed during the Pleistocene, and it must therefore be guarded, as it cannot be regenerated.

2

The nature of water flow
in the arid zones

Runoff

A considerable percentage of the arid zone rainfall is very localised,
limited both in space and in duration. Surface runoff, then, very seldom.
reaches the drainage system. This is particularly true in deserts of a low,
moderate relief. In other words, a small amount of rain, not exceeding a few
millimetres, may indeed start local runoff, but floods of regional dimensions
occur only a few times in a year (Planhol and Rognon, 1970).

Water flow in the arid zone is characterised by sudden floods; but even
sheet wash is very rapid, since it does not meet obstacles in the form of
vegetation or soil, particularly over granite slopes, if they are not fissured
and their surface is smooth. Nevertheless, this rapid sheet wash is hindered
by three factors: infiltration, blocking of channels, and the local nature of
the rainstorms.

Infiltration

Runoff is the excess of precipitation over infiltration and evaporation; it
starts when the amount of rainfall exceeds the amount of seepage for a
sufficiently long time to enable the filling of local water storages. The
volume of runoff increases when the intensity of rain and the slope gradient
increase, and it decreases when the surface becomes rougher and seepage
increases. The influence of slope length on runoff varies in relation to soil
permeability and rain intensity—it decreases on a long slope where soils
are permeable and rain intensity is small, whereas in reversed conditions,
that is when the long slope is made of non-resistant rock and the rain is of
great intensity, runoff values grow (Milthorpe, 1960). The total amount
of seepage from one single shower is influenced by the absorption capacity
of the soil-building layers. The capillary of the upper layer is of great
importance: a soil permeability of 36 mm (1·3 in)/hour under natural
conditions, increases to 78 mm (3 in)/hour after that same soil is ploughed.
When the surface is covered by thin crusts infiltration into the soil is delayed.
The initial rate of infiltration decreases as a result of the joint activity of two

processes: decrease of soil capillarity by the swelling of clay particles and by compression by raindrop impact. The latter process is a function of the size and ultimate speed of the raindrops. The rate of permeability is affected by the initial available moisture in the ground—it is greater in dry soil.

Since there are long intervals between, two desert floods (or between two flood periods, comprised of flood subperiods) the upper levels of flow in the alluvia are thoroughly drained. Another reason for this complete draining is the vegetation which draws to the last drop the moisture from the channel bed. The drained levels must, then, be refilled first, to make a new flood possible. Every hydrogram shows a sudden leap upwards at the beginning of flow, as it is preceded by a period of channel bed re-filling, which can be quite long. The flood then starts immediately with peak volumes, as the channel bed is already saturated.

Blockings

The more isolated floods are, the greater the number of blockings. The bed load carried by a certain flood is deposited at a certain point and forms a block against the succeeding flood, so that until the water in the channel rises to the level of the block, the flood cannot move on. When the block is broken through, an immense discharge may flow within a short time.

The local nature of storms

Rain is of a local nature and usually comes down in single storms. The storm's course and duration determine the extent of flow area, disregarding the form of the drainage basin over which the storm passes. An interesting course of this kind, for instance, was followed by the storm which passed over Sinai, the southern Negev and the Edom mountains in March 1966 (Schick, 1969, Fig. 2.1).

River flow and its intensity

River flow regime in the semi-arid zone does not differ, generally, from the regimes of the humid zones on which it borders: summer floods on the equatorial margins, winter or spring floods on the Mediterranean margins. But the general annual discharge per sq km is smaller than in the adjacent humid regions, usually less than 2 litre/sec/sq km and sometimes even less than 1 litre/sec/sq km. This is understandable when we consider the meagreness of total annual rainfall; even a big river like the Hwang Ho supplies only 1·9 litre/sec/sq km and the Rio Grande del Norte even less than 0·3 litre/sec/sq km (Guilcher, 1965). The picture is different when we examine a single flood in the arid zone itself; here we get very high values. Still, if we compare two big rivers, one in the humid zone and one

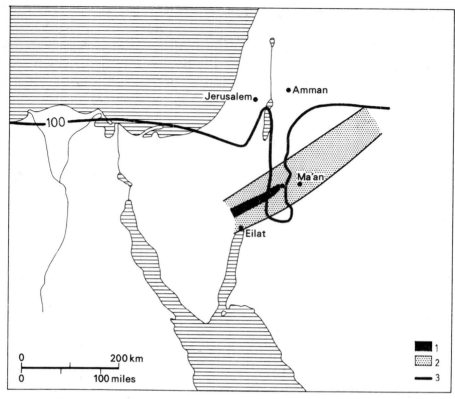

Fig. 2.1. The route of the storm of 11 March 1966 in the Near East. (Schick, 1969).
1. Region affected by heavy rainfall. 2. The path of the storm. 3. Mean annual isohyet
of 100 mm (4 in).

in the arid zone, the first will have supremacy. The maximum discharge of
the Hwang Ho in one of its greatest floods in 1843, was 36 000 cu m
(1 250 800 cu ft)/sec which is half of the Yang-Tse-Kiang's greatest flood
(Guilcher, *ibid*). The catastrophic nature of the Hwang-Ho flood was a
result of its unexpectedness (whereas in the humid zones, river floods occur
regularly every year), and because its valley and floodplain are densely
populated, so that there was damage to property and loss of life.

However, the feature which characterises flow in the arid zone is the
high turbidity values, which are the highest of their kind in all climatic and
hydrologic regions. The specific turbidity, defined as the weight of dry
load in suspension per cubic metre, varies from one river to another.
Still, it has values which are characteristic for the semi-arid climate; these
values increase from minima in the cold climate, through the temperate,
the equatorial and the Mediterranean, until maximum values are attained
in the semi-arid climate: 20 gr/cu m in the Siberian Yenisei, 20 kg/cu m
(1·2 lb/cu ft) in the Chaco river in northern Argentina, 78 kg/cu m (4·8 lb/

cu ft) in the little Colorado, 144 kg/cu m (8·93 lb/cu ft) in the Rio Grande del Norte. The highest value probably, belongs to the Hwang-Ho: 250 kg/cu m (15·6 lb/cu ft); that is, a quarter of the volume of a cubic metre of water is load in suspension. These high values of suspension content in its water may be attributed to the fact that the Hwang-Ho crosses the enormous loess deposits; cases of suspension load weighing 54 per cent of the total water weight of a given volume, have been recorded. The Missouri, on the other hand, though it has been called 'a big turbid river', carries only 2·7 kg suspension load per cubic metre (0·2 lb per cu ft).

There is a special relation between turbidity values and the amount of precipitation: in humid places, where precipitation exceeds 700 mm (27·5 in) a year, turbidity values are low, because of the uninterrupted cover of vegetation—usually forests or meadows—which prevents surface erosion. Where annual amounts are between 700 mm and 300 mm (27·5 and 11·5 in), particularly between 400 mm and 300 mm (15·5 and 11·5 in), that is, in the semi-arid zone, turbidity reaches the highest values, since slope erosion is fairly powerful, as a result of the nature of precipitation and the absence of continuous vegetation cover of the kind mentioned above. In arid regions, which receive less than 300 mm (11·5 in) rain, the water turbidity is also considerable, but taken in absolute terms (that is, the number of flood occurrences in one year or several years) erosion values are lower than in the region discussed before. It seems, then, that the semi-arid zone is probably the most eroded zone; some of its soils are still undergoing processes of erosion and destruction (Leopold, Miller and Wollman, 1964).

Flow in desert regions

Flow in desert regions—excluding allochthonic rivers, that is, rivers which though they cross the desert, originate outside it, and are not joined by tributaries along their desert section, which is clearly a section of loss—is only episodic, and river beds are normally dry. River flow in the desert is usually represented as catastrophic, since it destroys the river banks, obliterates traffic on the desert roads which cross the river beds, floods settlements which are built along the channel, or even on its bed, since the stream is almost the only source of water (for example, Feiran oasis in Sinai, which is built in the middle of the river bed), and drowns people who happen at the time to be crossing the river beds, which are among the most active anthropogeographic domains in the desert. It is said, with some truth, that more people have drowned in desert rivers than perished from thirst. Cars that take the risk and cross a flooded road are carried away by the water and the drivers pay with their lives for such attempts. Still, most floods occur on the desert margins, or else the source of their water is in the desert margins, where the streams come out of the high mountains in a concentrated flow of great intensity.

Examples of sudden floods of great, though shortlived intensity, exist in every desert location where it has been possible to measure the phenomenon, or even to observe it. One example that can be cited is that of a flood of 5 000 cu m (176 500 cu ft)/sec in the Saura basin in northern Sahara, at the foot of the southern Atlas mountains (Tricart and Cailleux, 1969). This flood, which covered an area of 22 000 sq km (8 494 sq miles), in the Colomb Be'char area, was the result of a rainstorm which yielded 81 mm (3·25 in) in three days. The length of the flooded channel course reached 8 000 km (5 000 miles), until it disappeared in the river bed as a result of evaporation and seepage. In less dramatic cases the length of the flooded channel reached only 300–400 km (185–250 miles). Fig. 2.2 illustrates the variability in flood dimensions in single occurrences.

Most of the drainage lines in the arid zone are endoreic, that is, they do not develop into a complete stream system that reaches the sea (de Martonne, 1927). In areas with a flat relief the streams do not reach any concentration basin, and a drainage system is non-existent. Such areas are areic, that is, they lack a drainage system.

The erosive activity of running water

The erosive activity of running water in this arid zone is conditioned, first of all, by the essential character of the flow: it is not regular, since the stream does not feed on spring water (though there are very rare exceptions, like the great spring of Ras el Ein on the Syrian-Turkish border, with a discharge of 2 000 million cu m (70 600 cu ft) a year, which forms a perennial river); flow is a result of rain only, so that the typical regime is that of an intermittent flow (Tricart and Cailleux, 1969).

It is difficult to make quantitative studies of flow amounts in the arid zone. As a rule the flow coefficient is expressed as the ratio between rainfall amounts (in cubic metres) and the discharge as measured along the basin's central drainage line (flow in cubic metres); when multiplied by 100, it gives the percentage from the total amount of rain received by the drainage basin. Still, while in the humid region this percentage can be calculated for the whole year round, in the arid zone separate calculations must be made, not only for the rainy season, but also for every single fall and its consequent flood. Measurements are made difficult also by the insufficiency of available data. The instruments which are used for measurement are not always reliable; the standard measuring instruments which are stationed in the river beds are not always prepared for unexpected discharges, which might even destroy the entire system, including the cemented river floors which serve as a basis for measurements (Nir, 1970). Nevertheless, all occurrences point to one conclusion: running water is the most important single geomorphologic factor in the formation of the arid, and particularly

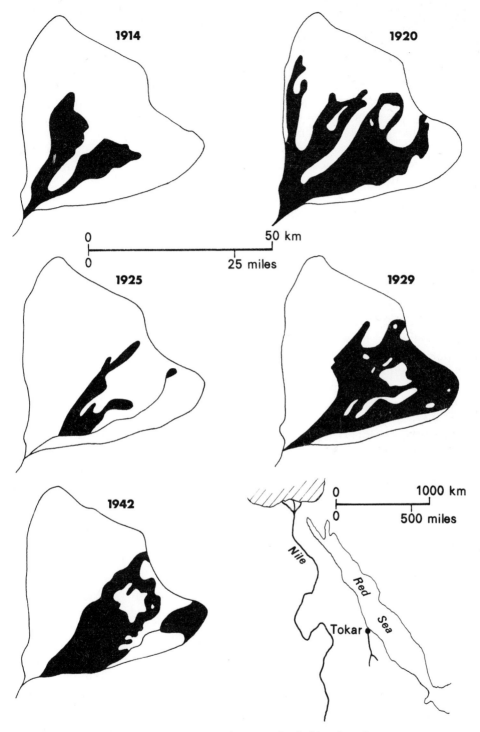

Fig. 2.2.   The Tokar delta in Sudan, showing areas flooded in selected years (Tothill, 1948).

27

the semi-arid, landforms. This importance of river flow derives from several secondary factors:

(a) *The absence of vegetation cover.* The amount of rain penetrating to the earth through the foliage cover depends, naturally, on the nature of this cover and its thickness. The smaller the total amount of rain, the greater the percentage of rainwater stopped by the foliage: of 25 mm (1 in) rain 55 per cent was absorbed by the foliage; of 5–12 mm (0·4–1 in) rain, 25 per cent was absorbed; only 10 per cent was absorbed from rain exceeding 40 mm (1·5 in) per day (Milthorpe, 1960, p. 10). This obstacle does not exist in the arid zone, and during the dry season not even in the semi-arid zone. Moreover, in the absence of vegetation roots to draw up water from the soil, the effectiveness of rain (for geomorphological processes) increases. The soil which is not protected by vegetation cover is directly affected by the mechanical action of falling raindrops on the surface.

(b) Arid soils are distinguished by their *high content of various minerals*, for instance, gypsum. Rainwater, if it attains an intensity of several millimetres in a short time, quickly saturates the upper layer of earth, and causes the swelling of minerals and rearranging of clays (Hillel, 1959); thus the whole depth of the saturated layer becomes impervious. This imperviousness prevents further seepage, thus creating favourable conditions for surface runoff. Such processes may not have been generated under conditions of less intense rains.

(c) *The great intensity of rain* is typical particularly of the semi-arid zone. A local flow, generated as explained above, is possible anywhere, only less frequently. Flow can occur in any part of the arid zone, if soil conditions permit. Lithological conditions are the decisive factor in determining the functioning of flow. Flow importance depends on the permeability of rocks; the generation of flow is possible only where the rate of rain water accumulation exceeds the rate of seepage through the rock. That is, for instance, the explanation for the arid appearance of sand and dunes, even in an area which is not definitely arid; in fact, any amount of rainfall will soak into the sand, barring the possibility of flow; one millimetre of rainwater will infiltrate into the sand to a depth of one centimetre, so that a deep layer of sand will absorb any amount of rain, and will prevent surface runoff. In all other types of rocks and soils some kind of flow system is formed; its value increasing in direct proportion to the decrease of rock permeability coefficient.

Flow coefficient, which is measured in the arid zone for every storm and the subsequent flood, shows great differences between individual cases. Goldschmidt and Jacobs (1958), claim that in the marginal Mediterranean region, such as the Judaea mountains, flow is not possible in a regional drainage channel, unless it received at least 300 mm (11·5 in) of rain during the rainy season, and even then, the runoff coefficient will be 1·5 per cent; but here the rocks are typically karstic—limestones and dolomites, which are very characteristic of the major part of the Mediterranean region. Evenari, Shanan and Tadmor (1963–64) measured in the northern Negev runoff coefficient values from 0·2 to 35 per cent; in Nahal Lavan, in the northern Negev, too, flow coefficients of 0·6 to 24·4 per cent were measured (Nir, 1970); here the soil is mostly chalky reg, anchored in loess soil, which very easily forms crusts. In southern Sahara, in the Ennedi mountain block, a flow coefficient of 55 per cent occurred in a small basin of 82 sq km (32 sq miles), after a storm that had shed 17 mm (0·6 in) rain in 32 minutes (Tricart and Cailleux, 1969). The Ennedi case occurred in an area with a sharp relief, which permitted concentration of the water flow. Usually, most flow coefficients in the semi-arid zone fluctuate between 7 and 15 per cent.

Flood dynamic

The dynamic of geomorphological processes is conditioned by their intermittent nature: the flood in the arid zone is characterised by the flood front, that is, the immediate large water volume, with the maximum discharge occurring at the beginning of the flood. The geomorphological importance of the large discharge is in its ability to carry the great amounts of waste material which accumulate on the valley slopes and on its floor during the dry seasons. The erosive power of the flood front is greater in a flood with a great hydrological cross-section (that is, a greater amount of water in a given length of time), than in a prolonged flood, which has a great discharge but is of long duration. The second type is more valuable for water supply. The flood front advances usually with great speed: velocities of 3–7 km (1·86–4·35 miles)/hour have been estimated by various researchers, and there were even estimations of 20 km (12·5 miles)/hour (Tricart and Cailleux, 1969). These velocities are similar to those of mountainous rivers in the temperate climate, which possess, as we know, great erosive power. As a result, the percentage of the load transported by the streams of the arid zone is high; we have already mentioned the high turbidity values measured in the rivers of this zone. However, we have very little information concerning bed load transportation, that is, the coarse material which is rolled along the river bed, and the greatest hydrologists doubt the possibility of making reliable measurements of this movement along the river (local cases can, of course, be measured, but it is impossible to draw an analogy about the general nature of this movement (Pardé 1951)). We know, however, the nature of this transportation: loads are being carried along with the maximum discharge of the flood. The components are not sorted at all, and all sizes of rock particles roll together, as long as the flood can carry them along. Then they are deposited on the river floor and remain there until the next flood. If this flood has a sufficient transporting power they are carried farther down the river course. Months, or even years, may pass between two occurrences of transportation. Any enterprise put up by man within the range of a desert stream's flow is in danger—perhaps not immediate—of destruction or damage by the catastrophic river transportation.

Groundwater resources

Water, especially groundwater, has been one of the basic elements of human ecology, everywhere and at all times. It is clear, therefore, that the availability of water, the knowledge of where to find it and how to make use of it, as well as the measures taken in order to store it and to ensure its availability, are of the first importance in the development of arid, or semi-arid countries. Surface water is almost non-existent in the arid zone, except in the immediate neighbourhood of springs, or in the few rivers that come from humid regions. In the last few years a great deal of information has been obtained about groundwater in the arid zone, both through activities concerning agricultural development and the growing needs of urbanisation and industrialisation, and last but not least, through its role as the forerunner of research and advanced experiments (Dixey, 1962).

The geological background for the existence of groundwater storages

The success of arid zone settlement depends on regular water supply. Shallow wells which supply only a few cubic metres per day are not regarded nowadays as a firm basis for advanced economy and a guarantee for a proper standard of living. The only dependable sources are, therefore, the

Plate 2. Windmill securing water supply, Manquehua, Chile (courtesy: the Director, Department of Geography, Hebrew University of Jerusalem).

allochthonic rivers which originate in some humid region, even if it is distant—the Nile, the Indus—or in nearby high mountains which are covered with snow—Sierra Nevada in California, the Pamir mountains in the Fergana valley, the Ebro in Spain, or the groundwater storages of considerable depth from which water can be drawn up. It is evident, then, that the term 'groundwater storage' has greatly changed its meaning during the last seventy years with the introduction of the gigantic drilling machines which go down to hundreds of metres, and of powerful pumps which can draw up water from such depths. The utilization of these storages depends on geological knowledge and on the economic ability to invest in such drillings. In most cases these enterprises are backed by public, national or international funds.

Rocks functioning as water storages

Groundwater is accumulated inside the rocks that form the earth's crust; the nature and distribution of these rocks determine the first condition for the possibility of water storage (Picard, 1952). The most important of the rocks that can contain water are the limestones (whose pore spaces comprise about 2–4 per cent of the total rock volume) and sandstones (whose pore spaces volume is about 20–40 per cent). The majority of the world's greatest springs—in North and South Africa and in the United States—come out of limestones; the reason for this is the existence in limestones of fissures, pore spaces and karstic caverns, which are absent in sandstones. Limestones, then, are paramount as underground water storages.

LIMESTONES AND DOLOMITES exist in various states of compression: the less compressed they are, the greater their water storing capacity; in this sense fossil coral reefs form better storages than compact and very fine-grained rocks. The humid conditions of the Pleistocene created latent karstic formations (caverns and subterranean conduits) which, though most of them are inactive today, still serve the southern Mediterranean region and North Africa as subterranean water conduits. These conduits are responsible for the existence of artesian water in many of the semi-arid regions. Lusaka and Broken Hill, in Zambia, receive hundreds of cubic metres per hour from Paleozoic dolomites; in Morocco the basic regional aquifer is the Paleozoic limestone of the Anti-Atlas; Ghardaia, one of the northern Sahara oases, gets its water from a Turonian limestone water storage, while other oases are located by springs coming out of Neogene lacustrine limestones (Touggourt); in Israel there are water storages in Cenomanian-Turonian limestones; the majority of groundwater sources of Anatolia, the Persian Gulf, South Australia and the Pacific Islands, come from limestones of various ages, from the Cambrian to the Pleistocene.

CONSOLIDATED SANDSTONES, particularly of the Paleozoic–Mesozoic continental formation, also form important water storages, especially in tectonic basin structures (see below). For instance, the Cretaceous sandstones of the great plains in the United States, Paleozoic rocks in the Andes, Karoo formation in South Africa, Albian sandstones in the Sahara, Nubian sandstones in the Arabian Peninsula and in northeastern Africa; the artesian formations in eastern Australia. These storages are the most important in the ancient masses of Gondwana. Here, artesian wells yield up to 1 000 cu m (35 300 cu ft) an hour.

UNCONSOLIDATED CLASTIC DEPOSITS, which fill internal basins, or overlie coastal plains, are quite recent (Tertiary–Quaternary). Their permeability is great and they form good water storages provided they are impregnated by impervious layers acting as aquicludes; such are gravels, sands and clays. Still, their discharges do not generally attain the dimensions of those from limestones and sandstones. Many springs come out of alluvial fans at the foot of the mountain ranges. These deposits maintain the underground irrigation system, such as the qanats which are common in the Near East piedmonts.

INTRUSIVE AND VOLCANIC ROCKS, which occupy about 30 per cent of the entire bedrock forming the earth's crust, are not important as water storages. Discharge from granitic rocks is particularly poor. In fractured metamorphic rocks storage conditions are better. Dikes and lava flows can act both as aquicludes and aquifers (the latter depending on their porosity), but vertical dikes are notable as water dams, and they form tiny water storages in stream beds; for instance, the wells in Nahal Feiran, in Sinai, are based on such dams (Issar and Eckstein, 1969).

CLAYS, MARLS AND SLATES are most important aquicludes; since they underlie and are interbedded with the aquifer rocks, they ensure the water flow on the storage floor.

Groundwater 'traps'

The mere existence of the right sort of rock does not, however, guarantee the formation of water storages; a certain tectonic condition might either encourage or hinder the formation of water storages. Picard (1952) distinguished several types of groundwater traps.

1. SEA-SHORE TRAPS, examples of which can be seen in the Sinai coast (Fig.2.3a). Here, water is accumulated in coastal-continental detrital deposits, such as gravels and sands. Groundwater is shallow and of small discharge. In cases where clastic aquifers alternate with aquicludes, richer and deeper groundwater sources are possible in coastal plains, too. Such cases exist in the coastal plain in Israel and in Morocco (Fig. 2.3a).

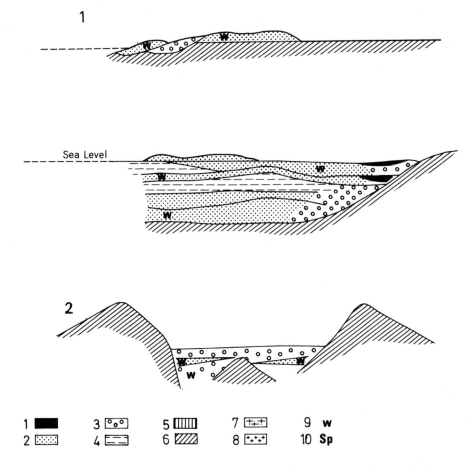

Fig. 2.3a. Water traps in the arid zone. 1. Loam. 2. Fine clastics. 3. Coarse clastics. 4. Clayey sediments. 5. Lake sediments. 6. Basement rocks. 7. Lavas. 8. Igneous rocks. 9. Water. 10. Spring (Picard, 1952).

2. TRAPS IN INTERMONTANE BASINS are found, for example, in the Andes or in the 'basin and range' region in western United States (Fig. 2.3a). Here too the storage is in young rocks of lacustrine deposits, surrounded by high mountains. Sometimes local artesian conditions may occur.

3. TRAPS IN A CRYSTALLINE AREA are scarce and hold little water. Small discharges of groundwater are found in waste pockets in the bed rock; examples can be seen in East Africa. Another form of trap is found in areas of lava flows underlain by volcanic tuff. Large discharges are possible in humid regions, but in the arid zone discharges are meagre. Springs are located at the base of lava flows, as in the Horan and in Yemen; an artesian

pressure may occur sometimes, as in the Columbia plateau in north-western United States (Fig. 2.3b).

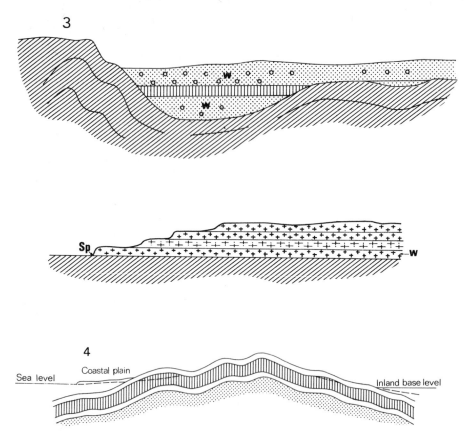

FIG. 2.3b. Water traps in the arid zone (continued from Fig. 2.3a).

4. TRAPS CREATED BY ROCK FOLDS vary in value, according to the nature of the folding. Folding of Alpine type does not generally provide favourable conditions for the formation of effective traps; moreover Alpine relief, that feeds directly on available surface water—snow, rivers, springs—is not common in most parts of the arid zone; on the other hand, more simple folding, which is found particularly in the semi-arid zone—the Middle East, the Saharan Atlas, the folded mountains of South Africa—enables the formation of traps at the foot of the mountains, for subterranean water that flows along the dip (Fig. 2.3b). In many Mediterranean countries this structure forms the basis for groundwater abundance in the coastal plain, at the foot of the mountains. Sometimes a considerable hydrostatic pressure is also present. Artesian wells may be found in synclinal regions, such as the Lebanon coast. The Algerian plateau, between the Saharan

Atlas and the Tell Atlas, may also be regarded as a great syncline, with the artesian springs of 'Chotts' producing between 500 to 1 200 cu m (17 650 to 42 360 cu ft)/hour.

5. BASINS OF REGIONAL DIMENSIONS exist in countries which were part of ancient Gondwanaland, and are located, in most cases, in ancient sandstones or limestones. These extensive basins are the classic areas of artesian traps in the Sahara, South Africa and Australia, and also in the Deccan. The basin structure sometimes terminates in huge cuestas, and the springs come out at the base of the scarp, as in the Sahara oases, or the oases at the base of scarps in the Arabian peninsula. The groundwater table is, sometimes, hundreds of metres deep, and the discharge may reach 300 cu m (10 500 cu ft)/hour.

6. RECENT FAULTING exposes, or brings nearer to the surface, water storages which normally exist at great depths; thus, in suitable lithological conditions, a drilling at the foot of a fault has a good chance of success. Sometimes, the faulting itself brings forth springs, as is the case of the important springs along the western margins of the Jordan rift valley (the Beth She'an springs, the Dead Sea springs).

Formation of groundwater reserves in the arid zone

The immediate climatic influences are perceived in the shallow groundwater tables; the deeper the groundwater table—especially if it is replenished from sources outside the arid zone through deep conduits—the smaller the immediate climatic influence, that is, seasonal fluctuations; and the chemical composition of the water is less influenced by conditions prevailing on the surface above the water table. It is true that the mineral content of groundwater increases in direct proportion to the depth of the water table. Still, in the arid zone mineral content is high even in shallow water tables, in comparison with shallow water tables in humid zones.

Groundwater can be replenished either by rain falling over the area, or by percolation from adjacent areas. The first case is important in the semi-arid zone, but only within the influence of the cold season rains; in summer rain regions evaporation is so intense that it practically annuls replenishment by seasonal rains. In all cases the determining factor in the amount of water required for replenishment is the surface cover. When the surface consists of exposed and fractured rocks, or of sands, replenishing may have a quantitative significance. In clayey soil runoff and evaporation exceed percolation. The wells of the great western ergs, in northwestern Sahara, may serve as examples: in spite of extreme arid conditions and the shallowness of the water table which feeds the oases, the mineral content is comparatively very low (300 mg of minerals per litre). The water quality in such cases varies according to the ratio of actual evaporation to the amount of rainwater penetrating the ground. When the amount of water penetrating the ground exceeds that which comes up to the surface, the quality of the water improves; when the case is reversed, the mineral percentage in the water rises. Sometimes, replenishing by a wadi (intermittent stream) is more important than direct replenishing by rain; this involves local and temporary base levels, such as stream floors or playas. According to certain estimates the ratio between direct replenishing by rain and replenishing by floods is 1 : 10 in favour of the streams. Replenishing by runoff is sometimes regarded as the only process that contributes substantially to groundwater replenishment in the arid zone (Dubief, 1963). Also, it is assumed that regular irrigation by river water—for instance, the Nile, the Indus and others—enriches groundwater in the neighbourhood of the irrigated lands.

The hydrologic cycle and water balance

The hydrologic cycle, that is, the system which transports the ocean water vapours to the continents and back to the ocean, is a well-known phenomenon which has been thoroughly explored. Various segments of this cycle may be governed in certain factors, such as the water content of soils and vegetation, and human activity in these two domains may cause modifications in the cycle. Knowledge concerning the conservation of such segments of the hydrologic cycle as can be controlled by man—soils and vegetation—is growing, but is not equally shared by all countries. It is particularly deficient in the countries of the arid zone which lack even basic knowledge; stations for the observation of flow regime have not yet been built; stream discharges have not yet been measured; and estimates of groundwater reserves do not yet exist.

Shallow groundwater is liable to dwindle as a result of wrong methods of soil cultivation, over grazing, or even excessive afforestation. A balance can be achieved by giving xerophytic plants, which require little water, priority over phreatophytic plants, which require large amounts of water. For instance (Wollman, 1962), the water consumed by phreatophytic woods in the southeast of the United States over an area of 3 sq km (1·16 sq mile), during one season of growth, could have supplied all the water needs of a town of 25 000 inhabitants. Therefore, the arid zone water problem is that of preference of needs: water exists, and man has to make optimal use of it for his needs; he may prefer a forest, or a wood, in a certain area, if it is required for his domestic fuel supply and for recreation, to the increase of population in that same area. Very important development enterprises have been carried out in this field of effective technological utilisation of groundwater.

3
Arid zone soils

The volume of the soil which is penetrated by plant roots is the sole source of water available to the plants directly. The size of this storage, its chemical and physical properties, its replenishing by rain water and draining up through evaporation, determine the amount of water available to the plants (Milthorpe, 1960). This definition of the soil is biological, and relates both to cultural and natural vegetation. In the arid zone the term 'soil' has also a geological meaning—that is, the earth's surface, even if it is solid bedrock, unsuitable for vegetation. Though soil in the geological sense exists everywhere, biologic soil, as defined above, is absent in most of the arid zone, and is not sufficiently developed even in the semi-arid zone, or else, it is extremely delicate. The reasons for this are the pedologic processes, which in the arid zone do not encourage chemical and organic weathering, and which are the only agents that can turn a geologic soil to a biologic soil.

Formation and classification of soils

Because of the climatic conditions which are marked by meagre humidity and high temperatures, rocks, as well as minerals, are not in advanced stages of weathering. Mechanical-physical breaking is dominant. The results are the undeveloped lithosols. These are classified in two groups: soils of deflation and soils of deposition (Aubert, 1962).

Soils of deflation

These are soils in which the fine grained components are absent, since they have been removed and carried away by the wind to adjacent areas. They are characteristic of the arid zone, where humidity is not sufficient to bind the fine grained components of the soil, and where there is no vegetation that could connect the clay particles. Soils of deflation are divided into three groups:

I. HAMADAS. These are plains, plateaus or mountains, made of exposed

37

bedrock, more or less dissected or rough. The soil components are the waste material of the underlying bedrock.

2. REG is a type of desert soil composed of gravel, usually very angular; this gravel, derived from resistant rocks (hard limestones, dolomites or flint), is anchored in a less resistant substructure (chalk, erosive limestone, marl). The resistant gravel is the waste material from more resistant strata which have already been disintegrated. The reg covers extensive erosional plains and slopes, when it is not transported far away from its place of formation.

3. SERIR is a type of desert soil composed of river pebbles that have been transported and deposited at their present site by river or flood flows. They may be found at great distances from their parent rock. Morphologically they constitute ancient river terraces, or ancient lake beaches.[1]

The gravels and pebbles which constitute these lithosols are often covered by desert varnish which is formed by the upward capillary movement of various oxides of the individual components. These soils, which are absolutely sterile for agricultural purposes, are confined to the deserts and do not exist in the semi-arid zone.

Soils of deposition

These consist mainly of two groups: soils deposited by wind, and soils deposited by running water of various regimes.

1. SOILS DEPOSITED BY WIND are primarily the sand accumulations, or ergs, which are found mainly on the margins of the desert; for instance, the ergs in northern Sahara. According to Dresch (1962) the original source of these accumulations are Pleistocene streams, which deposited the sand during the more humid periods of that era. Their present day movement, and also their present forms, are the results of transportation by wind. Another type of soil which has been transported by wind from its place of origin, and finally deposited by running water upon slopes, is the *loess*, which is deposited on the humid margins of the arid zone, that is, in the semi-arid zone. From here it is not carried by the wind further, since it is anchored to the surface by humidity, by crusts and also by the grasses which are the characteristic natural vegetation of the semi-arid zone.

2. SOILS DEPOSITED BY RUNNING WATER are usually found in closed basins which constitute the local base level of a certain drainage area—the *playas* (or *sabkhas*). Here, in the centre of the basin, are deposited minerals which

---

[1] The terms 'reg', 'serir' and 'hamada' are used by various authors for different purposes. An attempt is made here to classify soils types according to their components, and the names are borrowed from existing terms.

have been transported in solution, or come up to the surface by upward capillary movement; around them are deposited the clays which have been transported in suspension; and the outer ring is formed by sands which have been transported as bed load. These soils, then, are saline at the centre, semi-heavy in the middle, and light-sandy in the outer margins. These are the only soils with agricultural value in the extreme arid climate.

Soil development in the semi-arid zone

The conditions for soil formation in this zone, with an annual rainfall of 200–350 mm (7·8–13·8 in), irregularity of precipitation during the season and between seasons, and vegetation of the steppe type, produce brown prairie soils with a small organic content of only 2–3 per cent. The lower horizon contains lime, gypsum and soluble minerals.

This soil undergoes several more processes, through which it may acquire additional properties; for instance, hydromorphic processes produce heavier soils; lime crusts are formed, producing saline or gypseous soils. These modifications are related to the depth of groundwater table. A large portion of the lime crusts are not being formed nowadays, but are ancient deposits that had been covered by sands, alluvia or colluvia, and were subsequently exposed following the removal of the overlying layers. Hydromorphic soils are produced by an excess of water in the soil, by the

Plate 3. Badlands in Arizona (courtesy: S. H. Beaver).

upward capillary movement of groundwater, or by flooding. Soils of this type may sometimes be rich in organic matter, and form turf. In both cases, —crusts and hydromorphic soils—it is difficult to determine the share of presentday processes and that of paleogeographic processes.

Another group comprises the skeletal soils, which are soils only in the geologic sense; they are divided, according to the degree of rock disintegration, into lithosols, where the bedrock has not yet disintegrated and is not penetrated by tree roots; and regosols, where the bedrock is in an advanced stage of disintegration.'

The soils developing over loess deposits that have been transported by the wind, have not yet developed the profile which is characteristic of mature soils; the same applies to alluvial soils which are also young. These soils are from the agricultural point of view undeveloped.

Use and conservation of soils in the arid zone

There is a delicate balance in the arid zone between the soils and the factors of climate and vegetation, and any form of land use must take this balance into consideration. Extensive land use is suitable for the brown prairie soils, provided the right methods are employed. Slope cultivation must be done in terraces that take runoff into consideration. Extensive crops are very suitable for the natural properties of sand soils, particularly of those which are underlain by a clayey layer at a depth of about half a metre (18 in). Crop rotation is essential for preserving soil fertility. Erosion by wind is particularly dangerous, when the natural vegetation is removed to make way for cultural crops which do not cover the ground to the same extent. Extensive cattle breeding is also one of the forms of land use in the semi-arid zone (Fig. 13.2). But success in this field can only be guaranteed in areas that receive at least 150 mm (6 in) rainfall a year; and even then, when herds are too large for the pasture area, or are concentrated on limited areas, the consequence is a destruction of vegetation and danger of soil erosion. On the other hand, the improvement of pasture by the introduction of new species, such as leguminous plants, may at the same time increase the value of the soil.

Intensive agriculture is possible only with irrigation, either by inundation, a method that has been employed for thousands of years in the great river valleys, and is suitable for soils with a certain percentage of clay content, or by sprinkling, a method which has been developed only since the end of last century, and which is very suitable for a mechanised and controlled modern economy. Still, irrigation creates the problem of soil preservation against salination. All irrigation water contains a certain percentage of minerals; as long as this percentage is low, there is no immediate danger to the crops in the field; but even small quantities accumulate in the course of many years, and only effective deep draining can prevent salination of soils through irrigation, particularly when the water used for irrigation has a relatively high mineral content. Many soils have thus been destroyed in the semi-arid zone in the course of thousands of years of irrigation. The solution lies in leaching the soil with large quantities of fresh water, in a well-drained area. Many lands in India and in Egypt have been saved by this method—leaching and effective draining; and this method was also employed to prepare for cultivation the saline lands of the lower Jordan valley. One of the key methods for preserving the effective minerals in the soil is crop rotation. In the absence of a better method, soil fertility can also be preserved by leaving the field fallow. But above all, it is important to conserve the soil against deflation by wind or erosion by running water. In many cases the delicate balance of soils and vegetation has been disturbed through turning pasturelands into ploughed fields; ploughing in the wrong season, perhaps even during a drought, cultivation with heavy agricultural machines which compress the soil surface, all these have caused great damage. Many projects for turning 'virgin land' into fertile agricultural land have failed because the very delicate balance of the semi-arid zone was not preserved.

4

Problems of vegetation

Absence of water jeopardises the survival of plants, or at least hinders their growth. Natural selection and evolution have given birth to vegetation species which can resist the hard conditions prevailing in the arid and semi-arid zones (Oppenheimer, 1960). Such plants can endure lack of humidity, and can also resist high temperatures and radiation, winds and sandstorms. Generally, this quality of resistance in such conditions is xerophytism, but not all plants are of this type. There are, for instance, plants which come into being only with the rain, and complete in few weeks an accelerated life cycle of germination, growing and seed production. Such plants evade arid conditions and do not possess the defensive mechanism against such conditions; they cannot, then, be termed xerophytic. Other plants, which are equipped with a special root system to tap water from great depths, do not enter into this category either. These plants waste great amounts of water, whereas the real zerophytic plants economise in water. The characteristics of the real xerophytic plants are:

1. a small number of leaves which tend to fold and shorten their circumference, or even to disappear absolutely;
2. the thickening of the protective tissue to maintain the water supply (*Rotama Roetam*, Oleander);
3. fibre coverage of stems;
4. development of root systems that can embrace a larger soil volume from which water can be drawn. The roots are superficial and spread sideways in accordance with the volume which receives water; for instance, on stream beds or floodplains.
5. downward penetration—the roots of *Acacia seyal* have been known to reach down to 30 m (100 ft) in some cases.

Certain researches have attached great importance to the share of dew in the plant's water supply (Milthorpe, 1960). Dew is defined as water drops created by condensing of fresh air water vapour, usually over surfaces that have been chilled during the night by back radiation from the ground. Two types of dew may thus be distinguished: (*a*) descending dew, when condensed water vapours descend from the atmosphere to the ground;

Plate 4. Horseman in the 'caatingas' of NE Brazil clad in leather for protection against spiny vegetation (courtesy: S. H. Beaver).

and (*b*) distillation, when water vapours ascend from the earth and condense on vegetation surfaces. The first phenomenon adds water to the plant; the second only transfers water from one section of the system to another and does not add anything.

5

Man's physiological reactions to the arid environment

Many parts of the arid zone, especially those within developed countries, have become, since the Second World War, centres of attraction for a growing population of a new type—an urban population with a relatively high level of technology, but lacking the tradition and habits of life suitable to the arid zone. Does this tide express only a temporary tendency which must end in failure, or can the arid countries be turned into permanently occupied areas?

Past attempts at colonisation in the arid countries ended many times in failure (Amiran, 1966). The reasons for these failures were irregular water supply, diseases attributed to the special climatic conditions, economic and social failures, deterioration of the organisation of settlements and of irrigation installations. In modern life not only proper dwelling and sanitation conditions are of importance; a strong motivation is also needed to bind man to his new environment. The physical conditions of life can be organised so as not to form an obstacle; the question is not of ability, but of the price that has to be paid for it.

The physical conditions that characterise the arid zone—intense sunshine and high air temperatures—impose on man a considerable heat burden which can be eased only by intense water vaporisation through the skin. Man needs about 12 litres (21 pints) of water a day; if perspiration takes out of the body 2 litres (3·5 pints) per hour, at least an equal amount must be restored to the blood circulation; an increased consumption of water is necessary also in order to make up for the loss of minerals. The body cannot always meet the demands and disturbances in the blood circulation might ensue.

Every gramme of water that evaporates totally takes from the body surface 0·58 calories (at a temperature of 24°C : 75°F). As the temperature of the surrounding air increases, the importance of perspiration increases too; it is the only way to disperse the heat from the body. Perspiration is mainly a combination of water and salt. People marching over 25 km (15·5 miles) in the Arava valley, in August 1959 (Zohar, 1962), lost 10 litres (17·5 pints) of water in six hours; in the country's interior they lost only 5–7 litres (8·5–12 pints) during the same length of time marching the same distance. The amount of water consumed by drinking does not affect the rate of perspiring. Perspiring cannot be avoided by not drinking. If man does not drink he will be dehydrated. A loss of 5 to 10 per cent of the total liquid content in the body may result in severe objective disturbances; a loss of over 10 per cent endangers various parts of the body, particularly the brain, and might even threaten life. An unlimited supply of water has, therefore, to be available for men who work and live under conditions of high temperatures. Thirst is the main urge for drinking, but the feeling of thirst lags behind the actual deficit of liquids. Making do with little water, as is the case with the Sinai Bedouins, for instance, might be the proper answer to desert conditions, but it results, in addition to the dangers mentioned above, in inertia, lack of energy, and inability to perform jobs which could be done if unlimited supplies of drinking water had been available. Insufficient drinking is probably also the reason for the many cases of stones in the kidneys which occur in the arid zones (Gilat, 1962). Diseases which are the result of heat are rare, and can be prevented almost absolutely. Essential parts of the body, particularly the head, must be protected from intense direct sunshine. There is a group of diseases which are the result of exertion, among them the rare disease of heatstroke, and the less rare case of exhaustion caused by heat. These can be prevented by controlling exertion and preserving the body heat balance by drinking.

Fatigue is fairly common in hot climates, but researches show that not only physical, but also emotional causes contribute to this phenomenon. There are two faces to the definition of fatigue: one is objective—the decline in working capacity or efficiency; the other is man's subjective feeling of exhaustion (Marcus, 1962). Here we refer to the objective case of 'pathologic' fatigue when work efficiency declines not as a result of additional efforts. Researches carried out during the Second World War, among Australian soldiers in New Guinea, where the climate is hot and humid, showed that though there was no deterioration in their physical condition, there was a decline in the execution of tasks, and cases of carelessness, loss of appetite, and fatigue were observed. It was concluded that the decline in efficiency was not a result of the heat, but of emotional causes, which lowered motivation: discontent arising from work and dwelling conditions, from lack of entertainment, from isolation and severance, from boredom, monotony of the environment, inactivity; some of these are the result not of climatic conditions, but of the novelty of the environment.

We may conclude that in the existing technological conditions combined with an economic effort to maintain proper environmental conditions, a hot and dry climate does not constitute a particular hindrance to human life; from that point of view, these climatic conditions are far more comfortable than those of the humid equatorial climate, which encourage the existence of micro-organisms.

PART TWO

The wide spectrum of man's reactions to nature's challenges

Introduction

It is to be assumed that the first efforts of adaptation to environment were made by man in the very beginning of his existence as *Homo habilis*; later on, the efforts were dedicated to the improvement of the initial conditions. As conditions became easier, it became also easier to adapt to them. Regions with particularly hard conditions, such as extreme cold, high altitude, dense vegetation, deserts, could not develop into great centres of population, because they could not provide the basic necessities of man: food, water, shelter.

Growing technological expertise increased the capacity of man to adapt to his environment. Since our concern here is with arid zones, this adaptation means first of all the discovery and preservation of water. The simplest methods—as exclusive measures for dealing with this problem—exist nowadays only in a few isolated backward societies, and it is to be expected that technological change will eventually reach them, too. Still, the methods for drawing and delivering water have remained essentially unchanged from early times: man's muscles serve as the sole source of energy in wider areas than might be expected; the method of water delivery in jars carried on the head exists not only in residual societies with low technological level, but is also widespread in the Middle East, in India and even in southern Europe. In these cases it seems to be a socio-economic problem. A technological solution may be more effective in shifting the boundaries of aridity than climatic changes in the global wind system. In some cases the boundary of aridity is a matter of only 3–4 metres (10–13 ft) separating the Spanish *huerta* in the valley from the *secano* above its banks. There is a radical difference between the respective modes of life in these two areas. An uncomplicated technological solution could change this situation, which might have been, at the time, a fundamental and decisive factor; but in order to affect such a change, change-mindedness, a developed mentality and economic approach, and not least, a social approach, are necessary; a dynamic attitude to relations between man and environment is required. Yet, strange as it might seem, even today, in the last third of the twentieth century, such an attitude is not common property; many of the populations who live near water resources do not make use of them.

Plate 5. 'Bahari'—an irrigation system using only human muscles. Plain of Yamuna, near Aligarh, Uttar Pradesh, India (courtesy: the Director, Department of Geography, Hebrew University of Jerusalem).

There is no doubt that an examination of man's relations with his environment 200, or even 100 years ago would show the distribution of sophisticated technologies for adaptation to be much more limited than it is today. Similarly, it is to be expected that in a few more decades technological advances will have become more widespread. Yet the historical dimension is not the only determining factor; even in ancient times societies who employed advanced technology for water utilisation, and those whose technological skill was poor, existed at the same time, though they might not have maintained neighbourly relations, and were not, perhaps, even aware of each other's existence. The nature of the relations between man and the elements of his environment may be taken as the expression of the total sum of knowledge and capacity applied to the struggle for a better future. In some cases the natural conditions are fully accepted as an unchangeable element so that only the approach to them may be improved; in others basic conditions are improved by methods based on traditional knowledge handed on by generations, sometimes indicating great social efforts; and again, ideas, means and expertise may be borrowed from distant sources. Sometimes it would appear that the most important element in developing natural resources is the energy to act and the effort to master environment; but the energy and the effort are only the final expressions of a long and thorough development. A technical solution alone cannot serve its end if it is not associated with a general outlook on what man's life should be like; the more humane such an outlook is, the greater the chances for the technical improvement to succeed.

6

The acceptance of conditions as they are

The number of societies that accept natural conditions and turn their efforts exclusively to adaptation and not to change, is diminishing, and such residual societies as remain are little more than museum exhibits. Yet, while a museum's purpose is to preserve certain exhibits, the enlightened human society must raise these residual societies and provide them with higher technological capacity. However, the problem is not merely technological. Solutions for technological problems are usually welcome, but the problems are social as well, since their solution involves also a profound change in the way of life.

A human society that exemplifies—not in isolated cases of individual villages or families—a way of life with maximum adaptation to environment, is the Bushmen of the Kalahari.

The Kalahari Bushmen were called by many researchers 'masters of survival' (Van der Post, 1962, Debenham, 1953). The authors cited describe the immense effort which the Bushman applies to hunting with his most simple tools—a spear and a bow—or to the building of his hut from boughs of trees, or of the enormous exertion put in drawing by inhaling from the sands of a stream bed, an amount of 2–3 litres (3·5–5 pints) of water, which suffices for one day. The account of drawing water from the stream alluvium is probably the most dramatic description of man's efforts to reach water:

Near the deepest excavation Bauxhau knelt down and dug into the sand to arm's length. Towards the end some moist sand but no water appeared. Then he took a tube almost five feet long made out of the stem of a bush with a soft core, wound about four inches of dry grass lightly around one end presumably to act as a kind of filter against the fine drift sand, inserted it into the hole and packed the sand back into it, stamping it down with his feet. He then took some empty ostrich egg-shells from Xhooxham and wedged them upright into the sand beside the tube, produced a little stick one end of which he inserted into the opening in the shell and the other into the corner of his mouth. Then he put his lips to the tube. For about two minutes he sucked mightily without any result. His broad shoulders heaved with the immense effort and sweat began to run like water down his back. But at last the miracle happened and so suddenly that Jeremiah gasped and I had an impulse loudly to cheer. A bubble of pure bright water came out of the corner of Bauxhau's mouth, clung to the little stick, and ran straight down its side into the shell without spilling one precious drop!

So it continued, faster and faster until shell after shell was filled, Bauxhau's whole being and strength joined in the single function of drawing water out of the sand and pumping it up into the light of day. Why he did not fall down with exhaustion I do not know. I tried to do it and though my shoulders are broad and my lungs good, I could not extract a single drop from the sand. (Van der Post, 1962).

The Bushman's exertions are magnificent, but it is still strange that in the second half of the twentieth century man has to put in so much physical effort in order to obtain a small amount of drinking water. A small water-pump, with a two-horsepower engine could draw from the Kalahari sands, which are saturated by rainwater (400 mm : 15·5 in a year), enough water to provide for a tribe of thirty people.

Plant-gathering provides 80 per cent of the Bushmen's food (Marshall, 1960), whereas hunting is a more limited source of food. However, eighty-five plant species and fifty-four animal species are considered by the Bushmen as edible (Lee and Vore, 1968); but of these, only ten kinds of animals have been seriously hunted for food.

The question is whether the unusual capacity is to be admired, or has the time arrived for making efforts to make the Bushmen's life easier. At a depth of 200–300 m (660–1 000 ft) beneath the Kalahari sands, there are rich storages of groundwater, which could be drawn up to the surface and used for changing radically the Bushmen's way of life, which is now a continuous struggle for the bare necessities: all his efforts can only bring him barely sufficient food and drinking water. It is certainly valuable to teach the ways for survival in emergencies; it is important to the commando soldier who might find himself in critical situations in the desert, or to an astronaut who might have to land unexpectedly in the desert; but a whole ethnic group should no longer be forced to dedicate all its resources to the mere search for food and water.

Inasmuch as man is an ecological creature, the Bushmen provide, indeed, the most impressive examples of an ecological creature who is entirely dependent on his environment.

7

Use of natural vegetation and available water; nomadism and dry farming

Nomadism

If water supply is insufficient for irrigation, the nomadic use of pasture-lands in the arid zone may be the only method for man's subsistence, though on a low standard of living. We should not forget that the standard of living of the arid zone farmer has risen above that of the nomad only in the last few decades. Pasture is also the best form of soil conservation, pro-vided a balance is maintained between the pasture area and the number of animals grazing on it; it is preferable to dry farming, which destroys the soil crusts by ploughing, thus exposing the soil to erosion by wind in years of drought or running water in times of flood. Of course, such an attitude is baseless if the tendency is to raise the standard of living of the semi-arid zone population; then nomadism cannot compete with the higher forms of land use. Yet, the point about nomadism—though it is probably vanishing any-how—is that it is not only a form of land use but also a special way of life, associated with family and tribal organisation and with economic organisa-tion, which maintains relations with the world beyond the boundaries of aridity. This way of life is associated with the growing of unirrigated crops in various proportions; sometimes the crops are only a marginal occupation and pasture the main one; there is also the extreme reverse case where pasture is marginal and the main occupation is crop-growing. Between these two extremes there is a whole range of transitions from the arid to the humid zone. This combination is characteristic of the semi-arid zone way of life.

Nomadism is an interesting example of adaptation to an unreliable and poor environment. Nomad peoples have lived in the arid regions in many continents, from early days (Amiran, 1965). The majority of them are herdsmen, some are hunters and gatherers, like the Australian Aborigines. Most of them live in the semi-arid zone, but some live in the arid zone proper, like the Tuaregs of central Sahara. Unlike herdsmen in humid regions, the arid zone herdsmen must be content with poor pastures and travel for long distances. Pasture routes are constantly changing because of the unreliable rains, whereas the mountain herdsman in the humid zone, for instance, has permanent meadows as well as regular routes and time-schedules. All this is almost non-existent in the arid zone; also, the dependence of the herdsman on water sources for watering the herds is absolute.

Nomadism is the best form of adaptation for an unindustrialised society in the semi-arid zone, in that it adjusts itself to conditions without trying to improve them. Such ethnic groups are considered by some (Amiran, 1965) to be a social anachronism of our time. Migration itself—that is, the range of migration—is a function of the degree of the region's aridity. The more arid the environment, the greater the migration and the wider its range. In a less arid region, where pastures are more or less reliable and water sources permanent, the distances of migration are smaller, and the routes more fixed. In nomadism we have a mode of life and a standard of living entirely dependent on the region's primary natural factor—precipitation.

Nomadism in Mauritania

One example of a sociopolitical framework which is almost wholly nomadic, though it also begins to show the effects of economic changes which follow the penetration of foreign capital and the exploitation of mineral resources, is Mauritania. Barring a narrow strip in the south of the country, along the Senegal river, where a sedentary population lives on agriculture based on summer rains and the river floods, all the state's population consists of nomadic herdsmen. This population pushed southward the original sedentary Negro population, which left its imprints over the surface in the form of stone huts and grain millstones. These nomads live not in the desert, but on its margins, especially in western Mauritania, on the Atlantic strip which is a little more humid. In the region which receives less than 100 mm (4 in) rain a year, there is no human life—no water holes, no caravan traffic. In the semi-arid region small nomadic groups migrate along routes that have been fixed by tradition, as well as by the balance between the various ethnic groups and the natural factors. The caravans move southward while the pasturelands are being exploited, and northward when the summer rains are spreading in that direction. During the cool season the daily range of migration in the quest of grass may reach a distance of 20 km (12 miles) from the water source for cattle, and of tens of kilometres for camels. In the spring, when temperatures rise, the herds have to be watered more often, and the hard work of pumping water from the well starts. The methods are very simple: water is drawn by dropping down a skin bottle, at the end of a rope; the bottle usually contains no more than 5–6 litres (9–10·5 pints). Sometimes the hole is drained dry and some hours must pass before the water on the bottom is replenished by seepage from the strata.

Water holes become the centres of activity during the dry season. If a dense system of water holes exists, intensive use of the pasture is possible; if not, wide pasture areas are left ungrazed. Anyhow, the landscape is organised concentrically with the water hole in the centre. The first ring of soil, in immediate proximity to the hole, is compressed hard during the humid season and becomes dusty during the dry season, with no vegetation growing on it. In the second ring, 1 km (1 100 yd) in diameter, there are still a few scattered trees; in the third ring the vegetation is not so much destroyed and is intersected by many paths leading to the waterhole. Use of pastures is imperfect. Is greater efficiency to be achieved by a larger number of waterholes, provided that ground water conditions are suitable? Boring the holes, as well as maintaining them and pumping the water, without employing human muscle, involves investments that must come from outside sources; on the other hand, a growing population and larger herds may create overgrazing, with the consequent complications. There is no conceivable solution for this way of life unless it undergoes a radical change, such as basing its economy on mining or industrialisation. Water supply alone is not sufficient to increase the value of pastures (Daveau, 1968).

Nomadism and modern standards of living

The easiest solution for the problem of nomadism seems to be in turning the nomad into a cattle breeder. Although such a solution concerns all nomads, it is the least considered. Perhaps such a way of life seems anachronistic and is not rated high on the social scale; it is equally true that it is hard to define the required improvements without destroying an existing closed social system.

Nevertheless, improvement of pasture conditions is the best way to use the pasture soils of oases, especially on the margins of the Sahara, although they constitute a poor source of subsistence (Bataillon, 1963). Abandoning the pasturelands will mean a retreat of the inhabited area, since a vacuum will be left behind the nomads. Yet any improvement or change may meet with opposition in the social domain, since the nomadic societies are amongst the most conservative and have the most strongly established hierarchies. Any innovation which is liable to create changes in the social structure is apt to raise antagonism, just because of the aversion to the social change which is bound to follow. The area which by its natural conditions is the most suitable for modernisation is the Sahel, on the southern boundary of Sahara; but its social problems, of the kind described above, are the most serious of their kind (cf. p. 93–100).

The sedentary population of the Sahel is very scarce except in areas along lake shores and river banks, where it is exceedingly dense. The density of the nomad population is no more than 1–3 per sq km. It is generally assumed that the cattle breeder's standard of living is identical to that of the farmer, when a family owns about fifty head. But the average in Niger surpasses this figure: forty head of cattle, 200 sheep and goats and 10 camels. In Mali the average is 100 sheep and goats, 100 cattle. Even the camel breeders of Timbuktu keep 5 to 10 camels and 30 to 40 sheep and goats, which is a much higher figure than is common in northern Sahara.

This is not incidental; the quality of pasture in southern Sahara is better than it is in the north; moreover, water sources for the herds are more available here—the Niger, or the waterholes which draw their water from the groundwater table, which is relatively near to the surface here.

Improvement of pasture conditions must include, first of all, concern for the welfare of the cattle: veterinary services and vaccinations are becoming established measures in the region. Marketing methods are still in need of improvement, which means a radical change in the cattle breeders' habits of thought, as well as improvement of pasture plants. The problem of water hole improvement, on the other hand, is not serious and there have been achievements in this field in recent years.

Water holes are not very deep, no more than 50 m (165 ft); and in very few cases 70 m (230 ft). The existing facilities for drawing water—lowering a bucket tied to a rope and pulling it up by man or animal power—make it difficult to draw up water from greater depths. The water holes belong to a certain family, or to a part of it. The distances between water holes range from 20 to 30 km (12 to 20 miles); but their discharges vary from one place to another: from 6 to 10 litres (10 to 18 pints)/sec, and in some cases up to 300 litres (66 gallons)/sec. A water hole of the last type may supply drinking water to 6 000 cattle and 50 000 sheep and goats. The holes were bored by the cattle breeders themselves, or by their slaves; during the French rule a few scores of holes were bored. Abundance of water in itself is not a determining factor, since there is the limitation of vegetation wealth; it is preferable to bore many holes of small discharges with small intervals between them, than put large investments into deep drillings, and attain large discharges. This, of course, is the reasoning if the pastoral economy is to be preserved, without turning herdsmen into cultivators.

Pasture lands may be improved in two ways: by improving the existing vegetation, or replacing it by a cultural vegetation. In both cases there are the handicaps of the population's habits and of possible rejection by the natural environment. There are more chances, it seems, of improving the existing vegetation, than of applying the alternatives mentioned above. Attempts have been made to introduce cultural plants, as well as to store hay; these experiments have been opposed by the population; the nomads of the Sahara margins are not acquainted with the practice of reaping the hay, and turning it to fodder for use during the dry season. Other attempts to preserve the pastures—by leaving certain areas ungrazed temporarily and returning to them after a while—proved more successful. It seems, then, that development is hindered not by the nomads' ignorance of methods for pasture conservation or by any 'natural' reason—but by reasons of education, tradition and persuasion. The most important thing, perhaps, for the introduction of changes is a proper motivation: in the absence of aspiration to a different standard of living, different products, different tools, there is no great motivation for change. The aid of the state can be only an impetus. The realisation of the change is in the hands of the nomads themselves.

Marketing methods for cattle, sheep and goats

It is assumed that the value of cattle increases by 7 to 10 per cent a year; though this increase is not guaranteed every year, because of droughts and diseases, it still constitutes a certain profit. There is no commercialisation consciousness with regard to the herds; they are very seldom sold and turned into cash. The sheep and goats are used in place of money and in barter. Cattle and camels constitute the substantial form of capital; as such, they are something to be increased, not to sell or to be used in trade; it is, then, possible that herds will grow by improved pastures and watering facilities, and yet the region will not sell or export more than it has done. Still, the first consideration in the modernisation of cattle breeding is the ability to market the animals. The camel has long ceased to be of any importance as a pack animal, especially in the northern Sahara; in the south it has retained some of its importance. Here the factor of distance is significant: the distance from the Sahel to the markets of Ghana or the Ivory Coast, is 1 500 km (930 miles). The Chad is still further removed. Only an improvement in cattle transportation can contribute towards the commercialisation of breeding.

Nomads and the modern state

The mere fact of the existence of boundaries, mostly provincial boundaries which became state boundaries when the colonial rule terminated, is opposed to the nomads' tradition of free migration; there are today tribes that live within the boundaries of two or three states. Moreover, the 'state' concept and nationalist feelings are familiar to sedentary populations, but they are foreign to nomads, who live in tribal frameworks. The nomads are a very unstable element; their movements cannot always be checked, and most of the new states are therefore trying to convert them into sedentary people and bind them to a defined area. The nomads are a minority in all the new states of northern and western Africa, excluding Mauritania, where they constitute a very large majority.

In the USSR the problem has been to organise the nomads' life on a collective basis, that is, the tribal framework has to be replaced by the collective (Dresch, 1956). Before the Revolution the Uzbeks had lived in oases, and the Turkmen and Kirgiz had been herdsmen. Their way of life had been tribal-feudal. The real problems in their case were the wish to change the economic forms of organisation. An effort has been made to draw the nomads out of their closed framework and fit them into an open economy. Thus, for instance, hay-growing cooperatives were formed. Collectivisation of breeding herds raised some difficulties. In fact, the change has taken the form of transhumance—that is, regular seasonal migration routes—rather than real sedentarisation. The habitations remain in the plains, but during the summer there is a seasonal migration to the mountains; after a while the plain dwellings become a village.

Plate 6. A typical scene from a dry-farming environment near Aligarh in the semi-arid part of Uttar Pradesh, India (courtesy: the Director, Department of Geography, Hebrew University of Jerusalem).

The traditional farmer of the semi-arid zone

The farmer of the desert margins

Unlike the desert agriculture which is limited to the oases (not only from lack of water, but mainly because of the limited area of agricultural soil), semi-arid zone agriculture can be maintained over wider areas. Here the limitations are not of soil—though soil is poor, yet it is of the brown prairie type, not lithosols or regosols as in the arid zone—but in available water. The lands are not separated by stretches of desert, but are continuous. Still, the main difference between the arid zone farmer and the farmer of the semi-arid zone (meaning the traditional farmer) is that while the oasis farmer has no expectations from rain—he may collect rain water on his roof, but generally this water is a curse rather than a blessing (Amiran, 1965)—the traditional farmer of the semi-arid zone, who has not made any arrangements to receive water from another source, and whose sole source for water is the rain, has to struggle with this unreliable factor of the regional rainfall, of which the variability in amount as well as in timing is sometimes critical to his crops. This kind of agriculture has been defined as 'hazard agriculture'.

The irregular character of rainfall, and the annual variation of rainfall

59

totals, also sometimes make cultivation critical, since there is a great danger of soil destruction by powerful rains or by ploughing, which turns the soil and lays it bare, without protective vegetation, to the raindrops and running water, or to the wind. The destruction of the original grass by turning the pastures into ploughed land, that is, by upsetting the natural balance between rain, soil and vegetation, resulted in many economic disasters for new colonisers who came to the 'virgin lands'—as the experience of northwestern United States or north Kazakhstan will testify.

Not long ago another danger existed: invasion of nomads from oases, as well as from the arid margins of the region. For them the region seemed a 'land of milk and honey'. Compared to their tough conditions, they found here a land of fields and orchards. A great part of the history of the desert margin countries is concerned with the farmer who tills his land and the herdsman who brings his herds to graze on it after the harvest. There are even examples from Palestine at the end of the nineteenth century, showing that the Bedouin considered himself the real master of the land, permitting the farmer to grow his winter crops for 5–6 months, when there was plenty of grass anyhow; after the harvest he returned to 'his' pasture lands (Nir, 1968). The symbiotic relationship of the herdsman and farmer is not always smooth, and many frictions and wars are the result of this duality in land use; the conflict between Cain and Abel was only the first of many.

The farmer in oases

Life in oases has been based, throughout recorded history, on intensive agriculture of irrigated crops. Most oases are located in the arid zone, or even in the extreme arid zone, and thus form pockets of intensive agriculture in an area where other forms of life are, at the most, nomadism along the oases margins (Amiran, 1965). Only some of the inhabitants of oases are farmers, and social relations are very interesting, when the nomad warrior is the landowner and the master, whereas the farmer is his slave, usually also of a different race (Tuaregs and Negroes in the Sahara; the Hausa and the Fulani in the Sahel).

The term oasis refers primarily to the natural oases, that is, areas with springs which permit vegetation and organic life to exist, which are essential for the development of soils. Such an oasis, then, holds not only a water storage—which without suitable soil would not form a basis for agriculture, and remain only a drinking station in the desert—but also soil that can be cultivated. This soil, since it is limited, is exploited to its utmost capacity (having regard to the available amount of water, of course). Where there are perennial rivers or streams fed by melted snow, as in central Asia, whole villages arise, based primarily on agriculture; but soon these giant villages turn into urban centres, since they attract passers-by; they become trade and caravan towns, selling oasis products to wayfarers. From here, it is but a short step to the beginnings of manufacture, first as workshops serving the caravans, and later for commerce and marketing.

In very few oases is water dealt with on a really primitive level, without improvement of pumping and irrigation devices. Even before the introduction of the modern engine and pump, there existed in these places water-raising installations which were perfected according to pre-nineteenth-century technology: for example, the qanats.

The development of oases depended first of all, on the natural conditions of soil and water; but geographical location is also a natural condition of great importance, since it may put a certain oasis in a central position with regard to various routes, while another may be left in a marginal position; on the other hand, an important oasis by force of its own attraction, would draw the roads towards itself. The distance between a certain oasis and important population centres was also a decisive factor in its marketing possibilities; even today, distances from large centres of population outside the arid zone to settlements in it are of the greatest importance, in spite of modern means of transportation which shorten the distance.

8

Traditional irrigation on different levels of technology

Efforts to overcome the scarcity of water by storing, delivering and diverting it from abundant sources toward regions that need it, have been made in the arid zone throughout historic times, and almost everywhere; but the degree of success of such enterprises and also their extent and importance have varied greatly. In times past the main source of energy was provided by men and animals, but within the last hundred years the use of engines which are not necessarily bound to a close-at-hand source of energy has given an immense fillip to the development of the various water enterprises. The mere fact that water moves by gravitation made possible the existence of large and important irrigation works in quite ancient times, though they were bound by topographical limitations dictated by gravitation; running water was, in fact, one of the very first sources of energy.

Traditional irrigation methods

It may be assumed that the history of irrigation in the semi-arid countries is as old as their culture; in fact, the culture of the semi-arid countries is a culture of irrigation in its various forms. Life in oases depends entirely on irrigation.[1]

Many of the agricultural methods in practice today have their origins in the distant past; many tools have remained unchanged since Biblical times, or perhaps, even earlier—the threshing sledge and the wooden plough—and many of the devices for drawing water which are still used today have probably retained their original shape of ancient times (Cantor, 1967). The water supply sources were groundwater, as well as surfacewater. The resources of the semi-arid world are more variegated than those of the true arid world.

---

[1] It must be noted that 'irrigation culture' is not confined to the arid zone, and is to be found beyond it (Cantor, 1967). But irrigation in the arid zone is peculiar in that it is the *only* way to maintain agriculture and food supply; in other countries where irrigation is practised, such as the monsoon countries, or even humid regions with a more or less short dry season, irrigation organises agriculture and produces larger yields, but is not indispensable as it is in the arid zone.

Plate 7.   A version of 'noria', near Aligarh, India (courtesy: the Director, Department of Geography, Hebrew University of Jerusalem).

Groundwater supply

The simplest method, perhaps, for catching groundwater from shallow water tables, is the *tamila* method, that is, boring shallow holes in the alluvium on the floor of a desert stream, and drawing up the water which after an interval fills these holes by its gravitational flow. For instance, in Wadi Ruha, in central Sinai, there are orchards which are based on tamila; the water of Ein Furtaga, a spring in eastern Sinai, seeps into the ground near the spring, but reappears through tamila for many kilometres. In fact, the 'sipping holes' of the Bushmen also exist in the river alluvia and may be regarded as a kind of tamila.

Drawing up water from holes and wells has always been the safest and the most widespread method for supplying water, as this source is usually permanent and stable; the only changes occurring during the last two generations have been in the depth of the holes which now reach deeper and more reliable water tables, and in the introduction of the engine in place of man or animal as a source of energy for drawing water.

Among the most ancient tools, which are still in wide use, is the *shaduf*— which is a perfected method in comparison to the simplest rope-and-bucket method. The shaduf consists of a bucket suspended from one end of a long lever which is lifted by man or animal; the amount of water drawn up by

63

Plate 8. A 'noria' (worked by camel) in western Morocco (courtesy: S. H. Beaver).

this device at each lift is extremely small. This method is widespread even to this day, in northern India. An improved form of the shaduf is the *saqia*; a horizontal cogwheel, set in motion by an animal, turns a perpendicular wheel with jugs or buckets round its circumference; by the turning of the wheel the containers are lowered to the water, filled and brought up; when they come up the water is poured into a ditch or a trough. This method in a further developed form exists also on rivers, where the energy of the stream is used for filling and drawing up the water; the wheel is called a *noria*. One of the biggest known norias is to be found in the Homs area in northern Syria. The noria was among the most perfect pieces of apparatus in the days preceding the introduction of the engine; some norias, which are still in use, draw about 3·5 cu m (122·5 cu ft)/sec (Reifenberg, 1950). There are 'modern' saqias, not wooden but made of iron, and moved not by animal, but by engine power; hundreds of such saqias are operated, for instance, in the Ebro valley in Spain (Nir, 1967).

Of all traditional methods for catching groundwater in use before the invention of the modern engine and pump, the most perfected and widespread, demanding a high organising capacity and large capital and labour resources, is one that taps groundwater from the waterbearing layers beneath the surface, and delivers it in tunnels for many kilometres to the point where it can be effectively used—these are the *qanats* (or *karez*, or *foggaras*).

64

This method was developed in central Asia and Iran, and from there it spread to other semi-arid regions which are adjacent to high mountains, mostly snow-covered, where water infiltrates into the underground water storages. But the dip of these water-bearing layers is quite steep; and they are located at a great depth under the centre of the intermontane basin, where the agricultural lands suitable for cultivation are usually located. The problem is, then, how to catch the water-bearing layer and deliver the water from there in a minimal gradient, so that it will reach the centre of the basin by gravitation. The most detailed research done on village life which is based on still active qanats, is probably that of Humlum (1959), on the village of Pirzada in central Afghanistan (Fig. 8.1).

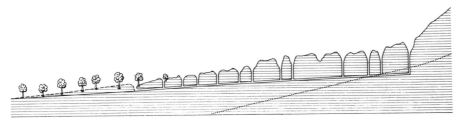

Fig. 8.1.   A cross-section of a typical qanat in Afghanistan (Humlum, 1959). The dotted line is the underground water-table.

The physical background of the *qanat* construction is the availability of mountains which get significant and regular rains, thereby guaranteeing the replenishing of groundwater storages. At the foot of the mountains there has to be cultivable soil with a land surface favourable for irrigation. This is the typical condition of many inner basins in central Asia, especially in Iran, Afghanistan and Turkestan. The first stage in qanat construction is to catch the groundwater table at the foot of the mountains, usually in the alluvial fan, by boring a shaft from the detrital surface down to the groundwater table; from there a tunnel with a minimal gradient enables the water to flow toward the centre of the agricultural areas. The earth is taken out of the tunnel through more shafts, which are spaced evenly; these shafts also serve to take out the earth that occasionally collapses and accumulates in the tunnel. The water thus flows slowly to the centre of the basin with no loss through evaporation (though there is no guarantee against loss through infiltration). The length of the qanats varies between several hundreds and several thousands of metres; some qanats even reach 25 km (15 miles). The qanat has to be cleaned out occasionally from earth that collapses into the tunnel; the openings of the shafts are, therefore, surrounded by mounds of earth taken out of the tunnel; the outward sign of the qanats is the row of these round mounds, 10–15 m (30–160 ft) apart. The number of qanats in Iran is estimated at 40 000 (Bémont, 1961). Though the digging of the tunnels is very expensive and requires high expertise which is handed down the generations, operating them is comparatively cheap; it is unlikely, however, that new qanats are still being constructed.

The qanats were generally distributed in the old world between China and Morocco; but it was undoubtedly the Persians who brought them to the distant places. It is known that the qanats of northern Sahara were built by experts from the Middle East (Papy, 1959). Some qanats exist in South America, where they were built by the Spanish.

The qanats in the Sahara have been the main technical devices used to maintain the oases in this arid region. There is information concerning the existence of foggaras in Marrakech in 1078 (Papy, 1959). The assumption is that the Barmaka tribe, which is still taking care of the foggaras, is a descendant of the Barmakids who came to the Sahara from Iran together with the Arab conquerors in the beginning of the ninth century.

The main area for qanats is in southern Algeria, at the foot of the Tademait plateau; most of the tunnels draw their water from the cretaceous sandstones. Another group of qanats is found in southern Morocco, in the southern Atlas mountains, in the oases of Tefilalet and in the Ouargla oasis.

During the 1960s the qanats of the Sahara were clearly deteriorating and the reasons for this are worth examining, since the qanat has been one of the most developed traditional methods for groundwater supply. Many of the oases which depend exclusively on the qanats for their existence are neglected; and there are numerous 'dead qanats'. The reasons for this are physical as well as social. Part of the palm groves die as a result of a fall in groundwater table, which in its turn is a result of a decline in the amount of rainfall. Another reason for abandoning the oases is the advance of the dunes. Attempts are made to stop the penetration of dunes by inserting palm twigs into the dune crest, but the sand rises and covers them; the dune thus becomes very high, until finally it penetrates into the oasis which is situated in a sort of bowl surrounded by sands. The sand also plugs the openings of the qanats.

The deterioration of the qanats increased when the French administration introduced social changes in the beginning of this century. The construction of the qanats and their maintenance was a collective enterprise; the qanat belonged to a group of owners, who took care of it and used part of its water; the work was done by slaves. The French administration abolished slavery; the slaves became hired labourers; the qanat owners were faced with expenses they had not incurred before. The qanat, instead of being a collective enterprise, became the property of those who had money to buy a share in it; thus the falahs who did not possess enough money to buy a share became tenants of the qanat owners. Those who had no lands abandoned the oasis. In many cases the qanat is not worth its expensive upkeep, deteriorates and is abandoned.

Artesian wells may constitute a better and more convenient substitute for the traditional qanat; the problem in this case is the basic investment.

The above remarks concerning the traditional ways of making use of ground water make it clear that except for the qanats, the methods were limited in the extent of their yields and areas. Deeper groundwater sources from water tables tens of metres beneath the surface were hardly used and were not considered 'natural resources'; groundwater at a depth of hundreds of metres has been untapped until the present century.

Surface water

In conditions of undeveloped technology surface water—that is, water running over the ground, such as rivers or overflow of springs—is more easily and conveniently available than groundwater; but there are limitations here, too. There is the physical limitation that the source of water must be situated at a higher level than the area to be irrigated, since water can be delivered to the field only by gravity (or by using one of the aforementioned devices for raising water, which are very limited in their discharges). The second and gravest limitation has always been the absence of perennial surfacewater in the arid zone. As has already been mentioned, there are absolutely no rivers in the arid zone, except for episodic streams or a few allochthonic rivers, such as the Nile, the Indus, the Euphrates and Tigris and the Rio Grande. Along these allochthonic rivers several of the world's most ancient irrigation cultures existed: the use of the Nile floods has been known since the third millennium BC; and the same applies to the basins of the Indus, the Euphrates and the Tigris. China constructed the 'Empire Canal', the length of which reached a thousand kilometres (620 miles), constituting one of the most magnificent engineering enterprises in the world, in the third century BC.

Nevertheless, there have been attempts to make use of the surface water of episodic streams on the desert margins, and they testify to man's immense efforts to obtain water in the arid zone. Diversion enterprises existed in the Near East in the second century BC, and during the entire classical period. Installations for diverting water and for irrigation, as well as important storing enterprises, can be found from the Negev to the foot of the Atlas mountains (Kedar, 1967). It is true that the lands irrigated from these sources were smaller than those irrigated by water from the great allochthonic rivers, but as a confrontation with environment, the effort made in these cases was no less important and no less decisive.

9
The technological revolution in irrigation and in the traditional farmer's life

The conditions described in the preceding chapter, of agriculture based exclusively on rainfall, or on irrigation methods on a low technological level, are disappearing from the arid zones; during the last few decades many countries have been improving their methods of irrigation. The old traditions have recently undergone greater changes than in thousands of years previously. The primary reason for this is the rapid spread of knowledge and technology resulting from modern communication facilities, and industrialisation, which attracts people from economically backward regions. At the same time, people who go on living in those regions expect living conditions to be improved. The backward regions have been losing population to regions where the economic demands of the people can be satisfied. People from drought and locust stricken regions can easily make a living in the humid region where manpower is in demand.

This part of the world has to compete nowadays with the standards of living that exist in the developed regions, or else accept desertion by its population. A farmer who has suffered from drought, locust plagues or crop diseases, will not hesitate to leave the region in search of a safer existence in adjacent areas where there is a shortage of labour. Only by raising the level of agricultural technology, thus reaching a standard of crop yield and economic profitability that can be compared with those of the more humid regions, can agriculture hope to exist in the semi-arid zone. However, the central problem here is not technological, but economic and political. What is the price to be paid for these improvements; in particular, what is the cost of basic investments, and who is to pay for them? Sometimes there are additional reasons, connected with politics or security, for preserving the population of these problem areas. The traditional situation cannot be allowed to persist. A choice must be made between progress and regression. In fact, the fate of the region depends today rather on political decisions than on natural conditions. The development of these semi-arid areas is possible, then, only in countries possessing the power for making political decisions, as well as sufficient monetary and technological means for pulling the region out of its backwardness and bringing it to a level at which its development may be guaranteed.

The first condition for improving agriculture in the semi-arid zone is its delivery from direct dependence on rainfall (Amiran, 1965). The region generally depends greatly on rains for replenishing its water storage—surface as well as groundwater storages—excepting such areas as get their water from allochthonic sources. Nevertheless, though such dependence exists in the long run, much can be done to annul the immediate dependence of the farmer by organising, capturing and making more efficient use of those same amounts of water that are received irregularly and unreliably; evapotranspiration, on the other hand, is an almost constant factor, because of the regular and unchanging character of insolation. Evapotranspiration can be counteracted by irrigation; the more advanced the semi-arid region is, the greater is the proportion of it under irrigation and the smaller is the share of dry-farming. Provided there is regular available irrigation, any semi-arid region, enjoying long hours of hot sunshine, can be turned from a backward to a flourishing economy; its high temperatures are an asset in contrast to the danger of frost, and the seasons of harvest and fruit-picking profit by the dry weather. There is thus every possibility of successful competition with humid regions. The prolonged season of growth in the subtropical semi-arid zone also turns it into the sole supplier of seasonal food (vegetables, fruits) for more humid and cooler regions. The early ripening gives the zone a monopoly in certain crops; this means, of course, passing from traditional agriculture of crops for local consumption, to a specialised market agriculture, run on economic lines; thus, for instance, the traditional wheat and barley are replaced by subtropical fruits, flowers and early ripening vegetables. However, agriculture is only one of the elements in a complex economic system, which requires the organisation of marketing, communication, transportation and research.

The effective use of groundwater: deep drilling

Among other technological revolutions of the nineteenth and particularly the twentieth century, which have influenced the arid zone, are the facilities that enable drilling to depths of hundreds of metres, as well as those which make possible the extraction of water from such depths; both improvements open up storages of groundwater which were previously unavailable to man. Deep drilling has provided arid plains and remote intermontane basins with the most reliable and controllable sources of water. Water is drawn up from drilled wells in accordance with defined needs, and there is no uncontrolled flow, as is the case with surface water. Nevertheless, highly technological extraction of groundwater depends on geological knowledge, technological expertise, sources of energy and, of course, large capital funds. We can agree with Karmon (1959), that the introduction of the water pump into the Mediterranean basin caused a revolution in the cultivation of light soils, and consequently in the regional

Plate 9. Well with leather buckets and a team of oxen in the Gangeric Plain of India (courtesy: S. H. Beaver).

economy and population. Karmon quotes as an example the Sharon area in the sandy coastal plain of Israel which had been neglected and devoid of economic significance until the end of the nineteenth century, as the light sandy soils were unsuitable for winter grain crops which need heavier soils with better moisture-retaining properties. On the other hand, the hilly relief did not permit irrigation by gravitation from the springs and the Yarkon river. The introduction of the diesel engine (before the availability of electric power) brought a change in the use of these soils; they were turned into citrus groves, a typical market crop, which contributed to a radical change in the economy of the region.

Of prime importance as sources of groundwater are the artesian basins (Dixey, 1966). The eastern parts of Australia—the inner plains of Queensland and New South Wales—have a claim to maximal use of artesian water. Water comes up naturally, or with encouragement, from depths from a few scores of metres to 1 400 m (4 500 ft), in the great Australian Basin, that extends over an area three times as large as France (Fig. 13.3).

Technological progress in the use of surface water

A really efficient use of surface water is possible only with new technology, capable of building dams that can withstand the enormous pressures of the water volumes that are held behind them, and of constructing cemented ditches that do not absorb most of the water before it can reach the fields to be irrigated.[1]

During the last century technological improvements have brought about a considerable and widespread development of great irrigation projects based on controlling surfacewater and connecting it to the irrigation system (and to the general water supply for urban and industrial needs as well), especially in countries possessing the socio-economic ability, such as the United States and the USSR, or even in developing countries like Egypt and Pakistan, that have regarded such projects as the decisive, and sometimes spectacular, step toward modernisation.

It is interesting to note, however, that it was in countries under the classical colonial rule that the first big dams were built—India (now Pakistan) and Egypt (Cantor, 1967).

Artificial filling of groundwater storages

The artificial replenishment of groundwater storages is becoming important for many countries in the arid zone, where water sources are available during a certain season when consumption is small, whereas there is a shortage of water during other seasons. The artificial filling permits the catchment of great amounts of water which without conservation would have been lost. This refers mainly to floodwater, that may carry millions of cubic metres which cannot be immediately used, but also to ordinary runoff and river flows during a certain season. This activity clearly depends on a detailed knowledge of the underground conditions, otherwise great expenditure may yield only meagre results.

Arid zones, and particularly semi-arid zones, are characterised by a short rainy season and prolonged dry seasons, with relatively great rain intensities (per day or per hour) during the rainy season. The problem is how to preserve the water efficiently. The simplest method, though it is not always effective, is that of open reservoirs, formed by dams the purpose of which is not to store the surfacewater for immediate use, but to insert it, through seepage, into the underground strata. It is relatively easy on the margins of the humid zone, where the stream load, particularly the suspen-

---

[1] For example, earthen ditches which existed in the Beth-She'an region in the years 1918–48 for delivering the Moda spring water to fields 12 km (7·5 miles) away, lost 40 per cent of their water through seepage and evaporation, since the ditch, in order to prevent friction and erosion, had been constructed with the smallest gradient possible, that is with long curves which contributed, of course, to this great loss of water (Nir, 1962), (Figs. 13·4, 13·5).

ded loads, is small; in cases of veritable floods in the arid zone, which carry large amounts of suspension load, there is danger that the capillaries in the soil and in the rock will be plugged, thus becoming impermeable and ceasing to function as conductors of water into the ground. This method for enriching groundwater has been used in several places in Israel, such as the Shiqma Project in the southern coastal plain and the Nahel Menashe Project near Hadera.

The best method, apparently, is to put the water down bore-holes. But the water must be from the ordinary river flow, not from floods, since the latter carry too much suspension load, that would block the capillaries in the bores.

The bore-holes in the Israel coastal plain feed the National Water Project during the summer, and together with the conduit's water join the national water supply system. In winter, on the other hand, when irrigation is less needed, but the Jordan's discharge increases, and the waterlevel in the Lake of Galilee rises, the surplus water is inserted through the bore-holes into the aquifer; in this way the water table in the bore-holes rises, and the water is preserved to be used during the dry season. Thus, the bore-hole serves as a conduit both ways—downward and upward, in accordance with the demands.

The most effective way to preserve water is underground. In open reservoirs the evaporation is great. If the reservoir is deep the expenses of construction are high, whereas in an underground basin, if it is sealed tightly so that water cannot escape from it, water preservation is optimal.

In the same way it is reasonable to control springs by drillings. While in its natural condition the spring water gushes out in accordance with the geohydraulic conditions, and its flow is uncontrollable by man, a series of drillings over the whole source permits the regulation of water discharge. This, for instance, is the case in the Rosh Ha'ain springs in the Sharon coastal plain, where the natural discharge reaches 180 million cu m (6 856 million cu ft) a year. Part of it formed swamps, another part flowed uneconomically to the sea; today these springs are caught by a series of drillings and their use is regulated.

Unconventional water sources: desalination

Saline water as a supply source belongs entirely to the era of modern technology; its development depends more on ecomic than on technological problems. A method for producing desalinated water which will cost no more than the water from conventional sources has yet to be discovered. Of course, where there is no fresh water at all, and there is an economic, military or scientific necessity to maintain a settlement, the economic calculation of desalination costs is not the determining factor.

Though in many of the arid zone countries the water potential has not yet been significantly developed—some exploit even less than 50 per cent of it, whereas others (e.g. Israel) reach 93 per cent (White, 1960)—the need to find unconventional water sources, that is, desalination of saline spring water or of sea water, is increasing. Already whole cities are living on desalinated water. What used to be an emergency measure—such as the desalination of sea water for the use of military units in waterless regions during the Second World War—is becoming today a necessary routine in waterless countries with an important economic potential, such as the Persian Gulf with its oil affluence. Desalination of water for urban and industrial needs in the arid zone has become in the last twenty years an ordinary procedure; the only remaining problem is that of the price to be charged for the water. It is evident that some activities can afford this price, while others cannot; what is possible for oil refineries or observation posts for satellites is not so for vegetable growers (especially if the vegetables are to be sold at a reasonable price at a distance of no more than a few hours' flight).

Research, development and technological systems for desalination of sea water or saline spring water have received much attention in recent years (Howe, 1962). Data show that though no revolutionary method for radically reducing expenses has been found, prices are nevertheless gradually declining. The prices of desalinated water in the arid regions are still four to ten times higher than prices of water from conventional sources in other arid regions, even though the supply of these too, requires considerable expenditure (e.g. for conduits and aqueducts, perhaps hundreds of kilometres long). For instance (Howe, 1962), 1 cu m (35·31 cu ft) from the desalination plants of Aruba in the Antilles costs $0·46 whereas the maximal price of water for urban uses in California is $0·10 per cu m (1960).

The great desalination plants, such as exist in Kuwait, California and Texas, produce about 5 000 cu m (176 000 cu ft) a day; other important plants are located in the Antilles, at the Dutch refineries, producing about 1 500 cu m (53 000 cu ft) a day. Other plants are located in the Persian Gulf and in South Africa and most of them are associated with the oil industry, or with cities, for example Gibraltar, Eilat, Port Etienne.

The mineral content of the water used for desalination is not uniform; the first distinction is between sea water and saline springs and groundwater; but the mineral content of sea water is also not consistent, and the seas which surround the arid zone have the greatest mineral content: the Persian Gulf and the Red Sea water contains 4·3 per cent minerals; the Mediterranean contains 3·94 per cent, while the Pacific and the Atlantic oceans contain only 3·6 per cent. The mineral content of saline springs and groundwater is scarcely more than one-third that of sea water: the salinity of the water in the Moroccan and Algerian oases ranges between 0·2 and 0·4 per cent, while in Karamoran, in central Asia, it is 2 per cent.

It is not intended to give a technical description of desalination processes here, but the main methods should be mentioned. The conventional methods, if they may be so called, are thermal: water is brought to boiling point, mineral-free vapours are collected and cooled. This method, though simple, requires much fuel and is not cheap; on the other hand, it has the advantage that it can be combined with a thermo-electric plant. It is also particularly suitable for sea water desalination, while other methods— chemical and electrolytic—are sensitive to the amount of salinity during the first stages of the process, and are therefore more suitable to saline water from sources other than sea water. The problem in the evaporation method is the cost, especially of the large amount of fuel required; there is a tendency, therefore, to use nuclear energy for evaporating the water. There are already about twenty different methods for water desalination and about twenty countries are occupied in developing methods and installations for this purpose.

Social and ecological consequences of the introduction of water into the arid zone

In contrast with the situation only a generation or so ago, the delivery of large amounts of water to the arid zone seems nowadays to be inevitable. Certainly such delivery is technically attainable; energy sources are more available than in the past, and topographical obstacles can be overcome. In many cases such delivery is also economic, since the value of water is measured not only by its direct product (agricultural crops, for instance), but also by the value of those same crops as a stimulator of more activities (commerce, industrialisation). For example, an artificial lake created by dam construction brings income in its capacity as a recreation centre, while its immediate purpose when it was planned was irrigation and electric power.

The delivery of water to arid regions is also necessary for the very existence of human society. As long as a radical solution for making the use of water for agricultural purposes more efficient remains undiscovered, water and not soil will remain the limiting factor of food supply to human society. Three tons of water are required today to produce 1 kg (2·25 lb) of wheat; in Texas it has been found that 200 tons of water are needed to produce 1 kg (2·25 lb) of beef (Wollman, 1962).

It is to be expected that any water installation of a considerable size will produce effects at the conduit headwaters as well as near its mouth; some influences may be felt also along its course.

The first result—in the case of a surface dam which is constructed for holding the water—is the inundation of the area assigned for the reservoir. This has an influence on the flora, which is destroyed, on the fauna and even on man; thus, for instance, Nasser Lake in Upper Egypt, required the evacuation and reconstruction of several villages; also the transference of the ancient Egyptian temple of Zimbel to an unsubmerged area. On the other hand, the mouths of the rivers that no longer carry their original volumes of water, are filling up with sand; the river course also changes.

This last detail is exemplified by the Yarkon river, which was fed at one time by the springs of Rosh Ha'ain; today all the water of these springs is taken by the national water project, and the river has become a drainage channel with all the pollution problems associated with it.

The sociological influences over man are, of course, more important. New villages are founded, and existing settlements may increase their population considerably.

The improvement of agriculture

The need to prove that agriculture can be maintained in the arid zone with sufficient irrigation is probably irrelevant today. The first thing to be examined is whether such economy can be self-contained or is to remain an eternal burden on the population, if the population is capable of carrying such a burden. On the other hand, the value of agriculture as a constructive element in the semi-arid zone is not to be despised. There is a place for specialised agriculture—the introduction of crops that during certain periods can be almost the sole source of supply, or crops that can attain unusual yields.

Israel specialises in many such branches of agriculture, both on a national scale and on an international commercial scale. On the national scale there are the early ripening vegetables of the Arava and the whole of the Jordan valley. On the international scale, there are the roses grown under plastic covers in the central and southern coastal plain; it is clear, however, that this branch which in two or three years has increased its sales from zero to several millions of dollars-worth a year, could not have developed without a wide organisation—credit for the great initial investment for the acquisition of vehicles, coordination with transport planes, market research and marketing in Europe. The roses, picked in the Lachish area in the morning, arrive before noon at the central packing shed in Lod airport, are sent to Europe the same night and distributed to wholesale markets in Europe's capital cities during the early hours of the next day; by early noon they can be obtained at florists in Paris, London, Frankfurt or Brussels. This detail may illustrate more than anything else the change of values that the semi-arid zone has experienced.

The supplying of water to compensate for the lack of moisture from rain is not the only way to overcome the basic natural factors. The challenge may be met from a quite different angle; the traditional crops may not be accepted as the only possible ones and an attempt may be made to introduce crops from similar ecological regions, thereby enlarging the choice of possible crops in the given area. Here too, science has created new possibilities: the selection of species, crossing and adapting them to the semi-arid zone conditions or to its soil, improvement of the soil by fertilisation, improvement of irrigation methods, protection of vegetation from cold or heat, protection from pests—all these have contributed to the

75

improvement of agriculture in this problematic region.

Among the basic conditions of any region, including the semi-arid region, are the distances to the various centres of man's activity: cultural, administrative, economic centres. The semi-arid zone has, obviously, been backward in the development of physical communications; actual distances have been doubled by logistic problems, such as shortage of food or water. Freight traffic to and from the region has been more problematic than in the temperate zone, where distance has only geographical meaning; the ecological distance has had no small part in setting the semi-arid zone apart from economic centres. Only in the case of important and expensive goods, such as spices and jewels, or when the purpose of traffic was beyond any economic calculation, as in cases of conquest marches did distance and travelling difficulties not carry their real weight.

The immense changes in transportation and in vehicles, which do not require today the same great number of stops for food and water along the road, have also brought about a change of values in the calculations for freight and crop transportation from the semi-arid zone to the developed humid zone. Suddenly, the semi-arid zone has been found to be actually very close to the temperate zone where food is needed for a dense population: the semi-arid zone has been proved, in fact, to be the vegetable garden of the temperate zone, especially during the winter season, and fresh

Plate 10. Vineyards in Upper Elqui, Chile (courtesy: the Director, Department of Geography, Hebrew University of Jerusalem).

Plate 11.   The Upper Ganga Canal near Aligarh. The Calcutta-Delhi highway passes the canal (courtesy: the Director, Department of Geography, Hebrew University of Jerusalem).

vegetables, fruits and flowers can be obtained at a distance of only a few hours' flight. These immense possibilities with all their implications for the types of crops grown in the semi-arid zone, have not yet been exhausted, and are only just beginning to develop.

10
Non-agricultural answers to the semi-arid zone challenges

While in the past most of the answers and solutions to the challenges of the arid zones concentrated on food supply—agriculture, cattle and sheep and goats husbandry—it seems that the future belongs to non-agricultural solutions. A parallel may be drawn between this problem and the general problem of world population growth; the future is in the development of non-agricultural branches. A constantly decreasing number of people are producing a constantly increasing amount of food, thanks to improving agricultural technology. It seems, therefore, that in the semi-arid zone too, the most profitable way of using its many resources—minerals in the ground, weather and insolation resources and vacant spaces—should be considered. More so, since improvements in technology and in water supply put the price of water as a first consideration, so that a maximal profitability and effective utilisation are to be aimed at. It is true that in the past too, there existed in the semi-arid zone non-agricultural forms of subsistence: garrison settlements, handicraft cities and cities of commerce and caravans flourished; but these are now undergoing a basic change of values.

The high economic value of water in non-agricultural sectors has been examined in two cities in the southwest of the United States (Garnsey and Wollman, 1963). This research has shown that the greatest waste of water is associated with agricultural-pastoral usage. The farmers comprise 9·6 per cent of the active population in Tucson, Arizona; they contribute 7·2 per cent of the total income, but are responsible for 46 per cent of the total water consumption in the area. The rest of the active population, occupied in industry and services, comprise 90·4 per cent of the total and contribute 92·8 per cent of the total income, but use only 53·9 per cent of the water resources. All this means that the agricultural-pastoral water returns a product value eleven times lower than the water put to other economic uses.

In Albuquerque, New Mexico, this is even more noticeable: 0·9 per cent of the active population is occupied in agriculture; their share in the total income is 0·2 per cent and they use 17·9 per cent of the water; to attain 1 per cent of the total income, agriculture will have to use 88 per cent of the water. Industry, on the other hand, employs 21·6 per cent of the active population, who contribute 21·3 per cent of the total income, using for this purpose 0·2 per cent of the water. To produce 1 per cent of the total income, industry needs 0·5 per cent of the total water resources, that is to say, 178 times less than agriculture.

Mines and minerals
Zonal characteristics

Mining is practised in all types of climate, from the polar region in northern Siberia to the jungles of Malaya on the equator. This section will discuss the problem of the extent to which this occupation might be given preference or be neglected in the semi-arid zone. This is associated with logistic problems of food and water supply, not only for human consumption, but also, and mainly, for the mining itself; the need for large amounts of water may determine whether certain minerals will be exploited or not; but besides this central problem, which does not encourage mining, the special mining conditions that may be favourable because of the arid nature of the region, must be examined (Hills, 1966).

Because of the meagre amounts of precipitation soluble minerals crystallise near or on the surface; in humid conditions these minerals would have dissolved and disappeared in the ground. Nitrates, which are formed during the decomposition of organic matter, remain on the surface in the arid zone; the example of the sodium nitrate deposits in the Chilean deserts, is probably the best illustration of the durability of soluble minerals in the arid environment, and of the economic importance of these large mineral concentrations.

Lacustrine deposits, such as ordinary salt, or gypsum, are also preserved on the surface in the arid zone, and excavating them is relatively easy. An interesting example is the salt deposits in southwestern Sahara, which are a residual of a Pleistocene lake; these are still exploited, and the salt is carried by camel caravan to the Sahel region. Gypsum, which is also a residual of lake deposits, is well preserved under arid climatic conditions, and it is mostly dug in open pits, as for instance at Machtesh Ramon in Israel.

A case of an organic deposit, which has been preserved in the arid zone, is the guano, the droppings of sea-birds, that is being deposited along the Peruvian coast; artificial islands have recently been built along the coastline in order to provide the birds with more nesting places. Absence of rain permits the accumulation and preservation of these deposits during the hatching seasons. A layer of 15–50 m (50–165 ft) of guano has been exploited (Hills, 1966), and deposits are still being formed.

Sometimes minerals are affected by local conditions in the arid zone: the fact that the groundwater table is usually very deep helps to preserve mineral concentrations from ancient and more humid times; that, for instance, is how the laterite soils, which store aluminium and ferrous oxides, are preserved.

Plate 12.   The township at Mount Goldsworthy, Western Australia (courtesy:
S. H. Beaver).

Despite human and technical difficulties, mining enterprises have always
existed in the arid zone if they were of economic or politico-military importance. One example is provided by the turquoise mines in Sinai in the time
of the Pharaohs. Armies consisting of hundreds of soldiers and slaves used
to depart to Sinai's turquoise deposits during the cooler, more humid
season; the ruins of ancient towns and temples, such as Abu Sarabit,
testify to the long life of this mining activity. There is no doubt that with
modern technology the obstacles of distance, transportation, climate and
food supply, can be overcome much more successfully than in the past.

Many new towns in the arid zone are based on mining. For instance,
Kalgoorlie and Broken Hill in Australia, both towns with several tens of
thousands of inhabitants, owe their existence to the mining of gold and
silver, respectively. It is clear that a town which bases its economy on a
single mineral industry is quite vulnerable; a crisis in the profitability of
the mines is liable to bring about a total collapse of the whole town; but
such a crisis usually depends on global market conditions, not on the arid
local conditions, though it is true that the desertion of such towns in the
arid zone is more drastic than is the case in the humid zone, where a town
can somehow manage to survive. The fact that 'ghost towns'—remnants of
the 'gold rush'—are to be found in the arid zone of the United States, and
not in the humid zone, is not accidental. In order to maintain a mining
town which produces valuable ores it is worth while to make enormous

efforts in organising food and water supply, as for instance, in the construction of the water reservoirs in Broken Hill, or the delivering of water to Kalgoorlie through a 500 km (310 miles) pipeline; the pipes bring irrigation to 400 000 hectares (1 000 000 acres) and provide drinking water for the small towns which are located along the route to Kalgoorlie (Hills, 1966).

Development works associated with mining in the arid zone involve large basic investments coming from public, government or private sources. But minerals of great economic value may not only contribute to the development of the arid region in which they are located, but also return to the state the investments put into them and constitute a source for general economic development. Mauritania, with its rich iron ores, is a good example; the oil of the Sahara states is also an impetus for an economic and demographic revolution extending far beyond the arid zone.

As a result of the mobile desalination installations, or the giant permanent installation where such is required, water shortage no longer constitutes a limitation on the development of mineral resources in coastal deserts (Meigs, 1969). The development of mining in coastal deserts depends only on the existence of minerals and on world markets. The best example is probably the oil industry, which is widespread in coastal deserts, as for instance, along the coasts of the Persian Gulf, Iran, Iraq, Saudi Arabia and Libya. Development lies not in production itself, but mainly in the erection of refineries, which can give employment to a large number of the popula-

Plate 13.   Iron ore mine from Mount Goldsworthy, Western Australia (courtesy: S. H. Beaver).

tion. The income from oil is put into developing and subsidising agriculture, public health and education.

Iron ores have also become an important development factor, in at least two places in coastal deserts, Mauritania and Western Australia.

The function of mines in the development of the arid zone in Chile

The historic function of mines in the colonisation of arid regions is best exemplified in the Chilean desert; though today the settlements there are not of the first importance to the state, a study of their development gives a notion on the economic importance and influence of this element.

For 150 years after the Spanish conquest Copiapo served as a station on the great north–south road (Bowman, 1924); its inhabitants made their living from pasturage and vine-growing. Gold was known to exist in small quantities in the stream beds, but the arid conditions prevailing in the Copiapo region did not provide the water for panning the alluvia. The introduction, at the beginning of the eighteenth century, of a different method—extraction from gold-bearing lodes, instead of panning the river gravels—raised Copiapo to a place of great importance; within few years it grew into a large settlement and attained the status of a town in 1744.

In the course of development, the climatic limitations of the region and its shortage of water became more pronounced. Mining required a great amount of water, but the undeveloped conditions of transportation prevented water delivery. Agriculture had to be limited in accordance with the limitations on water supply for irrigation. The concern with water supply was one of the chief problems of the town council.

Until the discovery of silver in the beginning of the nineteenth century, the Copiapo population amounted to about 4 000. After the discovery it grew to 12 000. Copper was extracted in Chile by the Spanish as early as the end of the seventeenth century, the main production being in Coquimbo and on a smaller scale, in Copiapo; yet extraction was not profitable unless the ores yielded more than 50 per cent pure metal. An increase in extraction occurred in the nineteenth century. The development of navigation and the introduction of steamships brought to the western coast of Chile ships which ensured its connection to Europe, as well as to the Far East. During the first half of the nineteenth century Copiapo's silver and copper production was increasing, while gold production was declining; in 1808 the region held 13 gold mines, 7 silver mines and 4 copper mines; in 1850 it had 8 gold mines, 235 silver mines and 24 copper mines; in 1866 gold was no longer produced, while there were 175 silver mines and 200 copper mines.

The dependence on natural conditions is illustrated instructively by the limitations on the transportation of the finished product. Transportation to Caldera port was maintained by mules. The year 1845 characterises the situation: since 1839 no ores had been transported to the port, since during the entire period there had been no rain and consequently no pasture for the pack-mules along their route. But in that year (1845) an abundance of rain in August created favourable conditions for growth, and in October the mule caravans took to the road—altogether 250 mules, who made the trip seven times between October and November. This was made possible by the grass that had grown in the months August–October. The entire distance between the coast and the mines was no more than 20 km (12 miles).

The main problem, then, was transportation; it is not by chance that the first railroad constructed in Chile was that connecting Copiapo to its port Caldera, a distance of 75 km (46 miles). It started operating in 1851, and was the first important railway in the whole of South America. Within three years Caldera grew from a settlement of fifty people to a town of 2 000. Yet towards the end of the century its population remained 2 000, because of a decline in copper production.

Urbanisation in the semi-arid zone: towns as traffic junctions

One of the oldest non-agricultural land uses in the marginal regions of the arid zone has been that of towns which were founded on traffic functions: caravan towns (though in many cases they were also connected with agricultural occupations). This type of settlement in the semi-arid zone has had general distribution in space, as well as in time; settlements engaged in providing the caravans with food and water, and serving as commercial centres at road junctions, have existed along all the roads of the arid old and new world for thousands of years. A town such as Palmyra in the Fertile Crescent in the Middle East, may be said to owe its existence to its location on the cross-roads of the north–south and east–west routes; towns like Bokhara and Samarkand in central Asia developed on the important routes from China to Europe and the Urals to Afghanistan; the settlements of northern, and particularly southern Sahara, such as Timbuktu and Gao, are also classical cities of this type.

This function has lost much of its value during recent decades, for the camel caravans have been replaced by motor-trucks which do not require so many stops along the route, so that distances between stops have increased.

The distinguishing characteristics of industrialisation

Even in the most traditional agricultural way of life there is a certain element, though usually limited, of handicraft: weaving, knitting, wood carving, building and tool-making. Yet there is a great difference between such occupations and a way of life based entirely on production for export. In the semi-arid zone, too, metal-working, jewellery-making, knitting and weaving have always been part of the life of the population. But these were not occupations on which a large population could base its existence. On the other hand, industrialisation, aiming at economic activities on a great scale, particularly for commercial and export production, did not exist in the arid zone to any considerable extent.

The existence of spacious empty areas combined with clear skies during the greater part of the year, attracted to the arid zone various modern military industries; ammunition factories, for instance, must be located at a great distance from centres of population, and the same applies to various nuclear plants; centres for testing nuclear weapons are all located in the arid zone (Reggan in northern Sahara, the Australian desert, Nevada desert and the steppes of Kazakhstan). Observation posts for artificial satellites also need clear skies and most of the telescopes following their movements are also located in the arid, or semi-arid zone. There is no doubt that such enterprises, though they do not support large populations, will, in time, become more valuable.

In recent years pioneer enterprises of modern technology have invaded the arid zone (Amiran, 1965). One reason for this invasion is that certain industries require large spaces, not only for buildings, but also adjacent spaces; there is also the advantage of a long dry season which enables storage, or even technical activity in the open. Vacant areas in the arid zone are cheap in comparison with prices in urban and industrialised regions; that is the case until the first plant is established; then, the mere fact of its existence, causes a rise in land prices. Moreover, the arid, or semi-arid climate, with its abundance of sunny days which encourages an out-door life, is preferred by the factory workers, who respond readily to the invitation to move from the humid zone to the zone with the more 'comfortable' climate. The rapid development of California, which is unmatched by any of the other states in the United States (or in the whole world, for that matter), is one of the proofs of this tendency. The development of the south of France, and even of the northern Negev, in Israel (the atomic pile in Demona) also point to this tendency as characteristic of the arid zone and are evidences of its importance.

This development is not an isolated case, of course; it is connected with a developed economic and technological infrastructure, which makes possible the lengthening of the lines of water and food supply, of electricity and telephone, of roads and railways, from the humid, densely populated regions into the arid zone.

A new function: recreation and tourism

Another branch which is non-agricultural but is dependent on a supply of fresh food and water, is tourism, whose share in the economy of the developed society is growing steadily; one company which organises recreation, the French Mediterranean Club, owns thirty-seven recreation villages

(1970) all over the world, most of them in the semi-arid zone. This tendency in the development of the arid zone is based, of course, on an economic situation which enables an increasing number of people to take more or less prolonged vacations; though it is true that in case of an economic crisis, this activity may be the first to collapse. The advantages of the arid zone are many, but the most important is its comparative proximity to large centres of population which suffer during the winter from hard climatic conditions, and whose inhabitants can reach the recreation centres in a few hours and for a reasonable price.

A special aspect of this is the increasing popularity of parts of the semi-arid zone for retirement; here a regular income earned by years of work elsewhere can be enjoyed in a warm and dry climate.

The coastal arid zone is particularly favoured; here, no other economic activity has as much chance of flourishing as recreation and tourism (Meigs, 1969). These areas serve not only the inland population which, however, at least in the arid zones of the old world, is unlikely to be very wealthy, but far more importantly, they attract the more affluent people from North America and central and northern Europe.

The arid zone coasts are distinguished by sandy beaches, and in many cases by a rich underwater coral world. Water supplies can be assured—at a price—and there are already towns living on desalinated ocean water. Development of the southern Mediterranean coast (including Morocco), which can offer the tourist interesting archaeological sites as well, is to be particularly expected. An example of an arid zone town which is developing mostly on the basis of recreation is Eilat. Sixty per cent of the town's drinking water supply is desalinated sea-water. Hotels have been built recently, and recreation provides jobs for 30 per cent of the active population. Another example is Baja California in Mexico, accessible by air or by road from the population centres of western United States; here the number of recreation centres is greater than in any other arid coast in the world. The small village of Ansanda has become a town of 50 000 inhabitants; its distance from the United States border is only 100 km (62 miles). Along with other services for tourists, a souvenir-making business has been developed. Further to the south, there are isolated hotels and motels, connected to the world outside by airlines or shipping.

Modern communications in the development of arid zones

The function of modern communication facilities, particularly aviation, in the development of the present world does not need to be explained here; there is no part of modern man's life where the influence of the quick and efficient means of communication is not felt; yet it must be noted that while aviation facilities in the temperate regions are important, but by no means exclusive, in the arid zone (and perhaps in other regions with difficult environmental conditions, such as the virgin forests, or the ice deserts) aviation permits the presence of population and its movement, which otherwise might not have been possible. It is clear that more than anything else, modern means of communication change the concepts of distance and time. Probably the most important example is the modern medical aid available by air in most developed arid countries. The 'flying doctor' service in Australia (Fig. 10.1) contributes to the safe existence of isolated farmsteads over the enormous spaces of the continental interior; instruction through broadcasting imparts elementary education to the children of these scattered settlements.

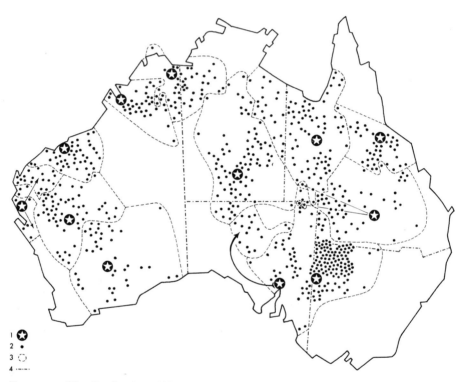

Fig. 10.1. The distribution of 'flying doctor' bases and contact points in the arid interior of Australia. 1. Base of flying doctor service. 2. Point served by the service. 3. Area served from a base. 4. State boundaries (Hills, 1966).

PART THREE

Human activity in different semi-arid environments

Introduction

The reaction of man to the challenge of natural basic conditions depends on many factors, among them ethnic, psychological, political and economic elements; their interrelations can be examined only in the regional complex, that is by studying a defined area where these elements are active in various combinations and intensities.

It is impossible to discuss here every place and combination of factors; the intention is to point out by a few detailed studies, the various complexes, the variety of applicable possibilities, and the lesson that man has learned from the continuous confrontation with his environment. There is scarcely a single region in which only one solution to the problem of aridity is adopted; it is, therefore, difficult to classify individual cases in unequivocal groups. The following discussion examines the cases selected in four different categories, each representing certain environmental conditions, though there are additional secondary conditions in each.

1. THE REGIONS ON THE MARGINS OF SEASONAL RAINS. Such regions exist on the borderlands of dry farming, and rely on rains as their main source for solving the aridity problem. These regions contain springs, wadis and groundwater sources that are drawn on for water supply; nevertheless, life in these regions is based mainly on rainwater. There are no great irrigation projects, no water is supplied from distant sources or from deep groundwater tables. Samples from the Sahel on the southern border of Sahara, the Karroo and the Veld in South Africa, part of Kazakhstan, the northern fringes of the Sahara and the Bedouin borderland in the Judaea desert, have been selected to represent this type of marginal area.

2. BASINS AT THE FOOT OF MOUNTAINS WITH ABUNDANT WATER OR SNOW. These regions are distinguished from the former in that independently of local rains they can get regular and considerable amounts of water from the adjacent mountains. This implies the development of a technology which can be a more decisive factor than the sources themselves. Afghanistan, the Fergana valley and the southwest of the United States illustrate various aspects of this type of region.

3. RELIANCE ON GROUND WATER AND SPRINGS. In this zone groundwater becomes the main factor among water supply sources (though rain and surface water still exist as secondary factors). The importance of technology increases, and in some cases the water resources can be controlled only by the most highly developed technology (according to present day standards). The great artesian basin of Australia and the Beth-She'an valley in Israel are examples of this type.

4. RIVER DAMMING is discussed here only in connection with the semi-arid zone, not with deserts; local and regional projects will be discussed, together with the changes caused by them. Northwestern Mexico, the semi-arid regions in India and Pakistan, and parts of the Ebro valley in Spain, represent this type.

Though this fourfold classification has been determined according to the degree of difficulty in establishing water control, it is not the intention to study the history of irrigation, or of its methods. Irrigation is only a part of the great complex of problems. The purpose is to observe beyond the generalisation the individual and local distinctions which are responsible for the success—or the failure—of the application of the general system to a definite locality. There is no one solution to all problems, and no lesson to be learned from the rule until its application to a definite limited area is examined.

11
Life on the margin of seasonal rains

It is hard to define this category as a separate unit among other possibilities of existence on the fringes of the arid zone, since any semi-arid region must be situated on the margin of seasonal rains. Nevertheless, it is possible to identify regions in which the agricultural economy consists mainly of unirrigated crops based on rains only, and also pastoral regions which depend, too, on rainfall as the only source of water for their grass. As a rule these regions do not have large nearby water sources, and so represent the correlation between rain variability and economic success or failure. At the same time, this extensive economy involves a certain form of life, and any economic change must be preceded by a social change.

The Sahel: cattle breeding as a basis for subsistence economy
Natural challenges

The southern borderlands of the Sahara are divided into two parts: the eastern—Sudan, which is connected with the north by the Nile so that the desert barrier is nullified; and the western—the Sahel, a strip of steppes and savannas, 200–300 km (125–185 miles) wide, constituting a part of several central and west African countries: Mauritania, Senegal, Mali, Nigeria, the Republic of Niger and Chad. The rainfall regime is that of the African monsoon; the monsoon winds move in a northeasterly direction from the coasts of Guinea, while the hot and dry harmattan winds move toward them in a southerly direction. The impact between the dry harmattan and the humid monsoon is expressed by a series of storms and tornadoes.

The instability of precipitation amounts which characterises the semi-arid zone is very evident here (Fig. 1.2). Thus, for instance, the average annual rainfall in Timbuktu is 240 mm (9·4 in) but intensities of 61 mm (2·4 in) in four hours have been measured; in Gao, with an annual average of 277 mm (11 in), 73 mm (3 in), or 26 per cent of the annual total, has fallen in a few hours. In Fort Lamy, Chad's capital, which is located south of the two former cities, and has an average of 643 mm (25 in) rain, an amount of 155 mm (6 in) was measured in two hours.

The dry and hot harmattan, which usually reaches the latitudes 12°–14°N in its southerly movement, sometimes reaches the coast of Guinea causing periods of drought. This wind has a negative influence on the water balance and dries up the soil.

The climatic conditions in themselves do not ensure easy and positive conditions of life; but the fact that most of the rivers that flow along the southern margins of the Sahel feed on summer rains which come down in the humid part of the continent, means favourable hydrographic conditions and regular supply of surface water during the flood season (Rodier, 1964). In the Sahel itself many parts of the hydrologic network are endoreic. The sudden precipitation, which is concentrated in space and time, creates a flood regime, and not every river channel is capable of coping with the sudden flushes of water. Only in the southern part of the region do the rivers flow the whole year round. The water disappears quickly because of the intense evaporation, that reaches its climax during the rainy season; hence the actual value of water in the region of summer rains is very limited in comparison with regions with a winter maximum. Furthermore, the structure of the region, which is composed of fossil quaternary dunes interspersed with layers of clay, leads to the formation of numerous small lakes that usually dry up toward the end of the rainy season. The percentage of runoff is low, reaching only 1·5 to 2·7 per cent.

This semi-arid zone is penetrated by several rivers whose source is in a region with plenty of rains—the Senegal, the Niger and the Chad and, in Sudan, the Nile. When these rivers enter the semi-arid zone, they acquire a braided pattern, with many branches, some of which fan out into the desert fringes and get lost. The reason for this is not only the present flood regime, but also the fact that some of the presentday ephemeral tributaries which are filled by floods, are residual from Pleistocene flow channels, in times when the Niger ended in an inland lake, and the presentday fan was actually the delta of that ancient lake. Today, the Niger channel consists of several branches and so does the Senegal river. For hundreds of kilometres the river flows in double branches. It is this detail of deserted channels which aids man in his irrigation projects; the channels for water delivery already exist, and it remains only to supply the water by constructing dams and diverting the flow into the old channels.

Out of the 270 000 sq km (105 000 sq miles) which is the total area of the Senegal drainage basin, 134 000 sq km (53 600 sq miles), that is half the area, are in the Sahel. From the point where the river enters this region it does not receive water; on the contrary, it loses water through splitting into two branches. The gradient of the channel is very moderate; along the river there are lakes which are, in fact, deserted ox-bows. At the peak of its flood the river carries nearly 5 000 cu m (6 500 cu yd)/sec, and brings to the Sahel 24 thousand million cu m (31 thousand million cu yd) a year. Part of this amount is diverted into a lake, in order to regulate the flow at times of minimun discharge; the dam near Richard Toll (Fig. 11.1) prevents the penetration of sea water into the lake and diverts the rising water into it. The lake has thus become an irrigation reservoir.

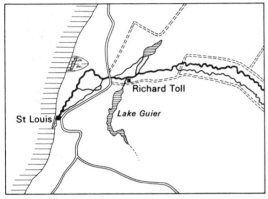

FIG. 11.1.   Northwestern Senegal.

The Niger is one of Africa's largest rivers. It is 4 200 km (2 665 miles) long and its drainage basin is about a million and a half sq km (600 000 sq miles). Its sources are in the humid monsoon region. The lakes along it in the Sahel region occupy a few hundred sq km; two of these are Faguibine and Debo (Fig. 11.2) whose area (and volume) change according to the season. The total area of the split branches—that is, the ancient delta—is 80 000 sq km (32 000 sq miles), and it forms a sort of amphibious area; this is the area of greatest economic potential in the Sahel.

The Niger's discharge on its entrance into the Sahel is about 70 thousand million cu m (91 thousand million cu yd) a year. On leaving the Sahel at the northern border, in the Timbuktu region, its channel carries only 38 thousand million cu m (49 thousand million cu yd)/year. The difference of

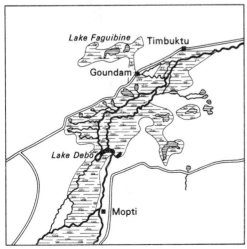

FIG. 11.2. The 'inner' delta of the Niger.

about 50 per cent is lost in the Sahel through percolation, evaporation and flow into deserted ox-bows and lakes. This water is not completely lost for human purposes; it can be recovered by not very deep drillings of no more than a few scores of metres.

The Chari river carries from a drainage basin of 600 000 sq km (240 000 sq miles) in central Africa, about 40 thousand million cu m (52 thousand million cu yd)/year (Rodier, 1964). This large discharge feeds the Chad lake. The upper reaches of the Logone, one of the Chad's main tributaries, approach the Benue river, which is a main tributary of the Niger in Nigeria. During its flood the Logone's water sometimes flows into the Benue; it is likely that the upper Logone may be captured by the Benue at that point. Yet, by constructing a dam the loss of water from the Logone may be prevented.

A considerable population lives on the fringes of the Sahara under these climatic and hydrographic conditions. The modes of life in this environment should, therefore, be examined.

Nomadism as a way of life

For many generations the commonest mode of life in the Sahel, particularly at great distances from the principal rivers, in the strip extending from Chad in the east to the Atlantic shores in the west, has been nomadism and subsistence on cattle breeding. Here live about 2 million people (1969), the majority of whom maintain a pastoral economy, keeping herds of about 12 million cattle and 23 million sheep and goats (Bremaud and Pagot, 1962).

The main problem is the quality of the pasture itself: in the temperate zone the pasture is of richer nutritive quality than in the natural pasture-lands of the Sahel. Unlike the temperate zone, in which most of the pasture is grass, mainly cereals, in the Sahel the pasture consists mainly of bushes and low trees. The quantity of food existing in the various plants is small, and it also varies according to the seasons:

	Rainy season (August)	*Beginning of dry season (December)*	*End of dry season (March)*
Humidity	75	48	9·5
Proteins	1·6	1·8	1·3
Fats	0·4	0·7	0·7
Non-nitrogenous substance	12	25	48·6
Cellulose	8·2	18·5	32
Minerals	2·3	5·4	7·3

(All figures are percentages.)

Of all the plant species that grow in the pasture area the cattle eat only 40 per cent; 36 per cent are never touched, and the remaining 24 per cent are eaten only when there are no other plants. It seems that an improvement of the pasture by introducing a new edible vegetation is essential for the development of grazing, in areas where this occupation is likely to constitute the economic basis of life in the future (Bataillon, 1963).

Another major problem is that of drinking water for the cattle. One animal may consume from 10 litres of water a day in September to 28 litres in May (17 to 47 pints). In order to supply water to the herds it is necessary to maintain drinking posts at intervals of no more than 20 km (12 miles); otherwise, the excessive walking would cause the cattle to lose weight. Wells of drinking water for the cattle existed during the French rule and have increased in number since independence was obtained. A well with a discharge of 30 cu m (39 cu yd)/hour can supply drinking water to 6 000 cattle or 30 000 sheep and goats. In order to ensure a regular supply of this amount it is necessary to drill modern wells, and this means investments surpassing the means of a single tribe. Government capital is therefore also put into these drillings. With the existing technological facilities greater quantities of water could be obtained, but the pastoral economy could not use the surplus. There is a distinct ethnic and historical division between cattle breeders and farmers, and the process of permanent settlement is even slower here than in the northern part of the arid strip in the Sahara or in the Syrian desert.

A regular supply of water solves the serious problem of thirst; it also creates new problems, previously unknown, though these too can be solved. The increasing number of water holes, and thus the gathering of a large number of cattle in one spot, facilitates the spread of cattle diseases; inoculation of the cattle against epidemics did not exist until 1965. Improvement in the water supply arrangements must, then, be combined with an organised veterinary service and an improvement of the sanitary conditions of the herds.

The movements of people and cattle are directed towards the water sources and the pasture areas. When the sources of drinking water become stabilised, the migration is also stabilised and turns into transhumance, that is, migration along fixed routes. Cattle-breeding is practised outside the Sahel also, on its humid margins, or along the rivers; but in these places it is not associated with nomadism, but with settlement; it is true though, that only certain tribes, such as the Puels (or Fulani) are occupied exclusively in cattle breeding.

Life in the Ferlo district

The Ferlo district is located in the northwest of Senegal, within the boundaries of the 300 mm (11·8 in) isohyet (Grenier, 1960). About 40 000 nomads live there on cattle-breeding; there is also land cultivation on a small scale during the rainy season, when millet is grown in selected regions (millet being, incidentally, a characteristic crop of the Sahel). There is, then, a regular annual migration from the 'winter life' area towards the millet fields near the Senegal river, as well as towards the water points which have been bored recently. Every such water hole becomes the core of a permanent small settlement, with schools, clinics, occasional visits from the veterinary doctor, and trading activity.

One of the development enterprises for raising the standard of living is located in this area on the Senegal river itself. The area is only a few metres above sea level, and at ebbtide sea water penetrates the river channel, for many kilometres. In the dry season the sea water bursts through an abandoned elbow and fills the Guier lake (Fig. 11.1). Dirt dams were built to prevent sea water from penetrating the freshwater lake; but every year this barrier was swept over by the rising water and had to be rebuilt to perform its function. A stronger dam was built in 1957, and the fresh river water is preserved in the lake. Since water for irrigation is guaranteed here, an experimental area of about 6 000 ha (15 000 acres) has been allotted for growing rice and groundnuts in an attempt to pass from an economy of subsistence cropping to commercial agriculture. However, it appears that by changing the ecological conditions new pests and animals have been introduced into the region. For instance, a certain type of duck which feeds on rice, has become naturalised here; flocks of starlings have devastated whole harvests. The entire project for settling 3 000 agricultural families in the region failed, as the yields were small (Church, 1961). Some experts have concluded from the Ferlo failure, perhaps too hastily, that nomadic cattle-breeding is the most suitable mode of life for the region and its inhabitants.

The inland delta of the Niger

Population density in the countries of the Niger's inland delta—Mali and the Republic of Niger—is 3·5 per sq km (1·2 per sq mile). The major part of the Sahel countries is practically uninhabited; the 30 000 sq km (12 000 sq mile) of the Niger's inland delta are inhabited by 13 persons per sq km (5 persons per sq mile) (Gourou, 1970). Agriculture is based on the river and its floods: rice fields, grazing on the margins of the channel in the floodplain, and fishing. The efforts of the authorities, colonial as well as the new independent governments, in the last twenty years, have been directed towards rice-growing for local consumption and cotton growing for outside markets.

The Goundam area (Grandet, 1958) is the centre of a district south of the Faguibine lake, west of Timbuktu (Fig. 11.2). Most of the area is fossil dunes rising over small drainage channels, some of which are tectonic. In the largest ones there are lakes which receive their water from the Niger floods. Around this centre there are numerous villages. The district extends over an area of 58 sq km (22 sq miles); its northern border touches on the desert; its southern part is a steppe; the rhythm of life is determined by the river which flows through it. This transitional area is also an ethnic meeting point between the Tuareg desert population and the Negro population of the south; the nomadic Fulani tribes also constitute an important element of the population. The climatic conditions are marginal for the existence of dry farming—an annual amount of 300 mm (11.5 in) rain, all of it falling in three months. There are in this district 140 000 inhabitants, 60 per cent of whom are sedentary farmers. Nomadism is the commonest mode of life here; the high percentage of permanent settlement is due to the additional element in the Sahel landscape—the Niger.

The inhabited part of the district is divided into seven subdistricts with a population density of between 21 and 35 per sq km (54–91 per sq mile), which is high compared with the general density of the country. A further examination of the ratio of population to the cultivated area shows very high densities of 200–433 per cultivated sq km (518–1 121 per sq mile). This density approaches that of developed agricultural countries in the temperate zone. It seems that despite the general demographical values indicating sparsity of population, there are actually very high densities concentrated on limited areas. This fact in itself is indicative of the low level of income per capita in the district.

Land use in the district is exemplified in the villaga of Qatana which divides its 183 ha (450 acres) in the following manner: sorghum, 100 ha (247 acres); millet 50 ha (124 acres); rice 25 ha (62 acres); and wheat 8 ha (20 acres). Division by the number of people shows an area of 0·5 ha (1·2 acres) per capita, which is really quite small.

The Kumare district, also in the Sahel region but nearer the town of Mopti (Fig. 11.2), also has a high density of population in relation to its cultivated area. The total area of the district is 200 000 sq km (12 500 sq miles); 21 000 inhabitants are scattered in this extensive area in ninety villages; some villages are big, consisting of 400–900 people each; but there are also quite a number of villages with no more than 100 inhabitants. Here, the escarpment which accompanies the Niger from the southeast recedes from the channel far eastward, leaving a spacious alluvial area which is inundated during the flood season, thus enabling settlements to exist. Fossil dunes arise here also over the alluvial plain, and the settlements are situated on these dunes, in order to escape the floods. The ethnic composition of the population includes both Fulani and Negroes. The agricultural economy takes on four forms: the breeding of cattle, sheep and goats, fishing, rice-growing, and dry farming.

There are here unique and most interesting interrelations between the fishermen and the rice growers, who are distinctly bound by the rhythm of the Niger floods. The breeders of cattle, sheep and goats maintain transhumance along the margins of the valley towards the desert. The range of transhumance today is some 300 km (186 sq miles) a year. Fishing in the

Niger, though not on a very large scale, occupies 1 500 people. The inland delta provides 2·8 per cent of Africa's fishery and the dried fish of Mopti reach the markets of the Ivory Coast on the Atlantic shores. The fishery here is second in importance only to that of Lake Chad, which provides 8 per cent of the total fish production.

The prospects for change

What are the prospects for the development of the Niger's inland delta which receives such vast quantities of water a year? During the 1950s a dam was constructed on the Niger near the town of Sansanding, diverting a part of the rising water into the abandoned meanders. In 1956 the governmental body 'The Niger Office' (which was French in the beginning and is now operated by the Mali government) prepared about 40 000 ha (100 000 acres) for cultivation under irrigation, of which 26 000 ha (64 000 acres) were allotted for rice and the remainder for cotton. The intention was to turn the Niger's inner delta into a 'second Nile', as regards amounts of water; this scheme is not far from being realised; but there are here grave ethnic and social problems. Water is available from the river's main tributaries, as well as from the groundwater table; still, in 1959 only 6·6 per cent of the total area of the inner delta was under crops (Gourou, 1970).

The fact of a sparse population distributed over the enormous area of the inner delta cannot be attributed to natural conditions only. Most of the delta's lands are fertile and easy to cultivate; the water flow regime does not present difficulties. It seems that human factors are at the roots of the delta's condition. The dozen or so tribes that are living in the delta have not been organised for generations in a national or any other framework. The delta area has been divided between various tribal and political spheres of influence. An attempt to unite the inhabitants under the rule of the Fulani leader Sheku Ahmade was made in the first half of the nineteenth century; but the delta has never been managed by any state which had the will and the power to organise hydraulic projects necessary for a large agricultural population. The invasion and conquest by the Fulani in the fifteenth century only damaged the delta's organisation; the original unorganised condition of the delta suited the nomadic cattle-breeding Fulani, since it made ideal pasture lands at times of drought in the north. Despite the attempts of the 'Niger Office', the population has remained poor and scarce; there is hardly any export of agricultural products or cattle.

Extension of the sown area in the virgin lands of Kazakhstan

The goal and the conditions

The north of Kazakhstan is located on the margins of the humid zone

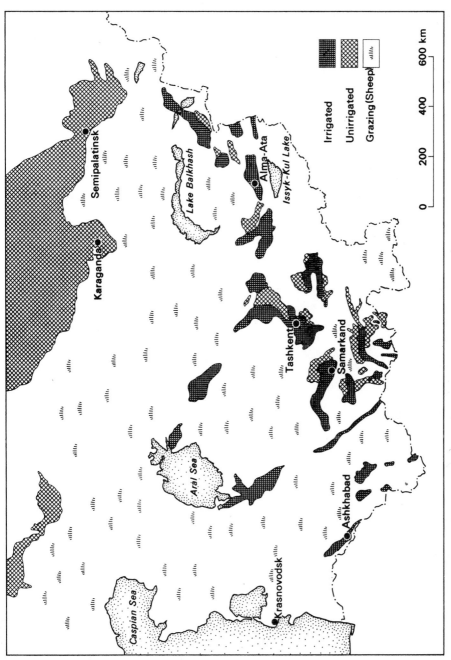

FIG. 11.3. Land use in Soviet Central Asia (Taaffe, 1962).

within the cereal strip which stretches along the southern boundaries of Siberia. The southern part of the Kazakh Republic is in the arid zone (Fig. 11.3), where permanent settlements can exist only by the meagre water sources. In regions such as this the idea of expanding the sown area by ploughing marginal areas formerly used only as pasture lands is not new; it has been carried out in various regions, such as southeastern Australia and the great plains of the United States. The Kazakhstan enterprise is exceptional in that it involved great national efforts and economic and political prestige; it embraced areas of unprecedented dimensions. Its aim was to expand dry farming in the most dubious climatic domain, the region on the borders of aridity.

The expansion of the sown areas has been posed as the goal of every Soviet Five Year Plan (Jackson, 1962). The first plan (1928–32) announced the enlargement of the sown area by 14 million ha (35 million acres), most of it in the steppe, which is in the Irano-Turanian ecological region. Other five year plans were interrupted by the Second World War. The 'virgin lands' plan (1954–56) was a sequel to the expansion projects; it was also very pretentious—it intended to turn 32 million ha (80 million acres) of pasture land into crop land. This immense project relied on new lines of communication, particularly the south Siberian railway; camps for the technicians and farmers were available; mechanisation was far more advanced than it had been in the former projects. The project's chances of success were undoubtedly quite good, during the first years of its existence, as long as the soil moisture and fertility had not been exhausted, and provided that no natural calamities, such as a prolonged drought, occurred. Cultivation was totally based on dry farming; the dangers threatening dry farming in this ecological region have already been described (p. 59–60).

Growth conditions are determined by the climate: only 100 to 130 days in the year are frost-free; on the other hand, the 250–350 mm (9·8–13·8 in) of rain, which is a marginal amount for unirrigated crops, fall mostly in the hot summer when evaporation is intense. Because of orographic conditions and also because of the region's position on the margins of the anticyclonic cell, the winds here are strong and erosion by wind is quite intensive.

The achievement and its lesson

Ploughing (in the 1954–56 project) was performed according to plan and even exceeded the area of the original project. A new infrastructure of new settlements was built in order to execute the project. The rate of ploughing slowed down after 1956, and from 1957 to 1960 only 4 million ha (10 million acres) were ploughed. In the years 1954–56 the newly ploughed lands of Kazakhstan occupied more than 55 per cent of the total virgin lands first ploughed in the USSR in those years—20 million ha (50 million acres) as against 15 million ha (37 million acres) in the Russian Republic.

Half of these were cultivated by collective villages and the other half by state farms. The sown land has thus grown in north Kazakhstan from 9·6 million ha (24 million acres) in 1953 to 28 million ha (70 million acres) in 1959. The checking of the rate of ploughing is understandable, since the reserves of pasture lands that can be turned into sown lands are not unlimited. The major part (64 per cent) of these sown lands was dedicated to wheat.

The planners were not oblivious of the danger of drought; the assumption was that if in the course of five years there were two good yields, two low yields and one medium, the expenditures would be bearable, and it would even be possible to attain wheat production with small expenditure. The assumption was that the average yield would be about 1 000 kg per ha (891 lb per acre).

The grain harvests in north Kazakhstan grew from 5·4 million tonnes in 1953 to 20 million tonnes in 1959. But an examination of the official statistics shows that yields failed to reach the expectations. The assumption that in five years two would be years of drought was proved wrong; in the years 1949–57 there were six years of drought out of eight. The first part of the summer—June—is usually drier than the remainder—July and August; it is, therefore, necessary to grow wheat that ripens later in the season; the trouble with this, however, is that if the crop is not reaped until the middle of September, it might be damaged by the first autumn frosts. A flexible attitude and improvisation are, therefore, necessary in dealing with methods of cultivation; and these are hard to attain in a centralised and planned economy.

The yields varied from one year to another, mainly, as mentioned before, because of the droughts. The best years were 1956 and 1958; the years 1954, 1955 and 1957 were most lean; the worst year was 1955, when only the extreme (the humid) edge of the region produced according to plan. The deficit between a good and a bad year was 25 million tonnes of grain; the significance of such a large deficit for the economic planning of the state is obvious. In addition to the drought damage, the early frost of September 1959 caused a loss of about 30 per cent of the entire harvest.

It seems that soil destruction by heavy storms is brought about, to a great extent, through deep ploughing. The method suggested (Jackson, 1962, p. 73) for improving the situation is deep ploughing once in four years, and superficial disc-ploughing in the rest. The superficial ploughing method is identical with the traditional method of the Mediterranean farmer who is also threatened by drought; he cultivates his land with a thin-bladed plough for that same reason.

Rotation of crops, fallow and the use of fertilisers were almost unknown in Kazakhstan; the method in use was that of leaving a field fallow for several years. But the system of compulsory quotas of production forced the farmer to plough many lands, and the fallow method was not strictly observed; soil erosion has greatly increased because of the continuous cultivation since 1956. The fallow, which creates a natural vegetation coverage over the soil, has been proved to be the best safeguard of soil structure against erosion; the absence of fallow resulted in the impoverishment of soil fertility.

The virgin lands project expanded the inhabited area in a region that previously had been the exclusive domain of nomadic herdsmen. During the two years between 1954 and 1956, 337 new governmental farms were founded in Kazakhstan alone. The cultivated area of a single farm was about 20 000 ha (50 000 acres), and the farms were ploughed mainly by a new ethnic element—youth from the western parts of the USSR, who must have had difficulties in adapting to the climatic conditions (as well as to the agricultural conditions of life).

As in any agricultural region on the margins of aridity that relies on unirrigated crops, here in Kazakhstan this type of agriculture is practised with a foreknowledge (or at least a forecast) of the risks. In contrast with conditions in the temperate humid regions, the length of the growing season here is limited by the drought, with the further addition, resulting from the continental climate, of early frosts; the risks are that much greater. The yields eventually became so low that high officials in the party administration, who were held responsible for the failure of the project, were discharged.

Consolidation and prospects of developments

In 1960 the virgin lands development area was declared as an independent administrative unit in the Kazakh Republic. It extends over 600 000 sq km (232 000 sq miles) and its population was about 3 million in 1963. The population growth-rate was one of the largest in the USSR: 80 per cent in the twenty years between 1945 and 1965, compared with 11 per cent in the whole of the USSR. Most of this growth came from the migration of a new population to the new lands. The greater part of the population is still agricultural, and only one-third is defined as urban (Karsten, 1963).

Agriculture is still based on unirrigated crops. Four-fifths are grain crops, mainly spring wheat: 13 million ha (32 million acres) in 1960. The mechanised equipment used for cultivation is quite impressive: about 200 000 tractors (with an average capacity of 15 horsepower), 77 000 harvesting and threshing machines. Yet this large region is still at the mercy of the rains, and drought has to be taken into account as a permanent element.

In the northeastern part of the region flows the river Irtish, which feeds on the snows of the Altai mountains. This river, and perhaps other Siberian rivers as well, might change the agricultural character of the region by turning unirrigated lands into irrigated lands which would be able to provide their harvest with more certainty than in the present situation and so to provide for further population.

Modern pastoral agriculture in an industrialised state: the South African Republic

Conditions and problems

Aridity and drought have always been the lot of South Africa, excluding the southwestern part which is temperate (Mediterranean) and the eastern mountainous part (Fig. A.7, 8). Likewise, the problem of water supply, for man as well as for animals and agricultural crops, has always been grave, although rich water storages have been found even in the arid zone. Three-quarters of the area of the South African Republic receives an average of less than 500 mm (19·5 in) rain a year, and the greater part of the rain (85 per cent) falls in summer, when it is less agriculturally efficient than cool-season rain. The runoff rates are estimated to be 6 per cent, which is quite a high figure; though more than half of this amount reaches the arid zone through river channels, surface configuration, the absence of dams and the absence of soils near the channels, offer little encouragement to the development of irrigation. The Mediterranean region in the southwest of the country has always maintained an agriculture based almost exclusively on irrigation, whereas the semi-arid zone has remained the domain of stock breeding. The high degree of industrialisation in the Republic, and the development of mines as early as the middle of the nineteenth century, which provide a high standard of living, have brought this country to a state where agriculture is not the only source of livelihood, as is the case, for instance, in our former sample, the Sahel. Gold alone constitutes 45 per cent of the total exports of the country (Gourou, 1970). The various branches of agriculture, especially the pastoral branch, support only 13 per cent of the entire white population, and about half the non-white population. Still, agriculture is not to be dismissed easily: the 'merino' wool, over a million tonnes a year, is the second important item, after gold, in the export list.

Only some 15 per cent of the country's cultivated lands are in the arid zone (*sensu lato*). Cattle-breeding is concentrated mainly in the humid zone, and only 20 per cent of it is in the semi-arid zone (although these 20 per cent mean 2·5 million heads). On the other hand, 45 per cent of the sheep and goats are maintained in the semi-arid zone. 10 per cent of the rural population from all tribes live in the arid zone (Talbot, 1961).

The landscape of the semi-arid zone is an open steppe with scattered acacias. Originally this region held a rich and varied fauna: gemsbok, antelopes, wild ostrich and springbok were abundant. It became the Hottentots' pasturelands for cattle and sheep and goats as early as the fourteenth century. Even today pastoral land use remains dominant; irrigation and ploughland are scarce. Yet, by more intensive use of the pasture and mainly by the possibilities of export, the breeding of sheep and goats has become a definite market-oriented occupation.

The penetration of civilisation into the semi-arid zone in the present century

The expansion of modern civilisation into the region with no available surfacewater had to await the drilling machine which could reach down to the available groundwater sources, and the windmill which could draw this water up to the surface; in the conditions of communication at the end of the nineteenth century, devoid of the conventional sources of energy, such as coal or wood, the wind was the first source of energy that could be harnessed. Windmills, scarcely heard of in 1890, have since the beginning of the present century, become a characteristic feature of the landscape.

In the north of the arid zone, near the border of Botswana, in Bushmanland, groundwater is found at a depth of 30–60 metres (95–190 ft). Until 1908 there had been no permanent settlement in this region, and the lands used to be leased for a year as pasturelands, or otherwise sold in auction on a twenty-one year lease. Only when large drilling machines were brought to the region—as a result of a military expedition during the First World War to German South-West Africa—and the internal combustion motor replaced the steam engine, was it possible to make large numbers of relatively inexpensive drillings; yet even in 1960 there were still farms with no permanent source of water, and 9 000-ha (22 000-acre) farms with only one drilled well.

After the First World War attention was paid to the settlements in the southern part of the Kalahari (within the boundaries of South Africa), along the rivers Molopo and Kuruman. The new farmers usually did not possess great means; the state has therefore subsidised their settlements and even sold them lands on a sixty-five year repayment basis, at 1 per cent interest. The conditions in the beginning were difficult and the absence of capital was a drawback. The distance to the city where they could sell their products—mainly skins, or wool—was seven to ten days of driving in ox-carts; this is too great a distance for delivering fresh products, and the farms could not, of course, be based on dairy cattle. The lack of capital also prevented the installation of modern pumps for drawing water from the wells which had been dug by the government in every farm it sold. Water was drawn from depths of 60 m (190 ft) by hand-pumps. In such extreme cases it was necessary to employ four people in pumping from morning to evening in order to provide drinking water for 2 000 sheep and goats, and 180 cattle, which constituted the size of the model farm.

Though nature has supplied surfacewater in niggardly fashion, the natural pasture in its original state was luxuriant and sufficed even for preparing hay against the dry season. The supplying of hay may be regarded as the beginning of intensification and commercialisation. The fruit of the various acacias is edible and can be preserved for use in the dry season. Yet this kind of pastureland was not without its problems: because of the geological background of crystalline rocks, the phosphatic content in the soil is meagre. The absence of this element from the fodder caused a paralytic disease which spread as an epidemic in the early years of the present century, and almost ruined cattle-breeding in the region. When the cause of the disease was discovered, the regular supply of phosphates cured the cattle that were brought into the region. When the Mafeking–Kimberley railway was constructed, the cattle ranches within a distance of 150 km (95 miles) from the railway line began to develop rapidly; 150 km is the distance that cattle can cross on their own feet without suffering too much. New roads and the organisation of motor transport, characterising the second quarter of the present century, made possible the breeding of dairy cattle; by 1930 the regional transportation services were delivering over 5 million litres (1 million gallons) of cream a year from these parts. Today cattle are kept mainly for milk, to serve the large and not too distant urban population. Cattle breeding is no longer an entirely separate activity—as it is, for instance, in the Sahel—but an essential link in a developed economic system which under certain circumstances is prepared to pull it out of a temporary crisis.

The danger of the desert's advance

Droughts have always contributed to the semi-arid landscape in South Africa. Yet according to data gathered since 1860, they are no commoner today than they were a hundred years ago. On the other hand the desert threatens to advance into the region: the floral landscape is more desertlike than it was. The reasons are cultural, not climatic; the replacing of the original fauna by a more sedentary and homogeneous fauna—cattle, sheep and goats—has utterly changed the ecological situation.

The findings of the Committee for Examining the Desert's Advance, in 1952, asserted that the misuse of the natural flora in the South African 'veld' had reached such dimensions that the vegetation was everywhere receding and being replaced by an invading xerophitic vegetation with species that are less nutritive for pasture purposes (Fig. 11.4).

One of the explanations seeks the cause in the psychological background of the settlers. The Boers, South Africa's European settlers, were actually of a half-nomadic character; turning into settled farmers did not suit their mentality. The Boers' grazing system was defined even in 1908 as unsystematic, primitive and wasteful (Agnew, 1959).

The pasturelands were exploited beyond their capacity, first around the ponds and watering points. During the floods the grazing routes became channels of running water and severe erosion began. Nevertheless, as long as migration persisted, there existed also a sort of seasonal rest; with the stabilisation of pastures this seasonal rest also ceased, and as a result the vegetation coverage has been thinning; the capacity for absorbing water has decreased, erosion has increased, and it is apparent that the desert is advancing. The primary damage was caused early in the present century,

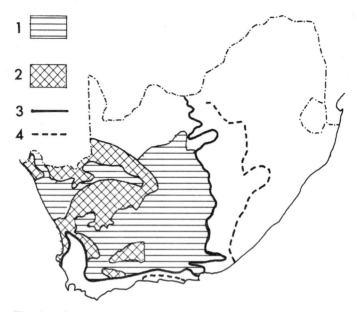

Fig. 11.4. The desert's advance in South Africa (Talbot, 1961). 1. Land below the critical stage of desertification. 2. Culturally induced desert. 3. Limit of arid zone. 4. Isohyet of 500 mm (19·5 in).

when lands were only given on lease and could not be held permanently by the cattle-breeders; when the farmers purchased the lands they began to improve soil conservation. The committee for Examining the Advance of the Desert points out the danger of 'the formation of a large uninhabited South African desert'. This warning of the committee has influenced the urgency of the assistance provided by the South African Government; since 1960 this Government has been much concerned with improving the conditions of agriculture. The veterinary service has been extended, and there is also a more general concern, since it has suddenly become evident that the agricultural regions which were formerly the economic centres of the country, have become its problem areas. Government assistance, direct as well as indirect, has become an obvious procedure.

Intensification of farming by irrigation and improved pasture

The influence of regular irrigation (as distinct from well-drilling for the supply of water for animals) on the development of the arid regions in South Africa is limited. Only part of the river courses lies in the arid zone, and their flow is highly irregular; for instance, for some years the Orange did not flow at all at its lower reaches. The surface runoff may range between 0·8 and 35 per cent, so that these sources are not reliable for water supply. Moreover, because of the comparatively recent uplift of the continent, the rivers occupy narrow valleys and cut down deeply below the plateau surface. Their gradients are very steep and there is but little extent of fertile alluvial land on the floors of the valleys. Groundwater, too, is not of the best quality; 30 per cent of it has a mineral content that makes it unfit for animals. The high percentage of load material in the rivers poses problems for reservoir construction. Along the Orange river there are only 32 000 ha (80 000 acres), which can be irrigated, and these, too, only in small scattered plots. About 500 000 ha (1 235 000 acres) are irrigated today in the whole of South Africa, but more than half of this area lies in the humid zone.

One of the most important changes that has contributed to the improvement of the pastures is the division into fenced plots. Certain areas are left to rest from grazing and their food reserves are replenished thereby. Every such fenced area possesses its own water source for the herds, so that they do not have to migrate far. This efficient organisation has increased the grazing capacity of some areas by 30 per cent in an ordinary year, and by an appreciable percentage, even in a dry year. Fencing has thus removed the necessity for seasonal migration; it has also prevented the spreading of diseases by migration. Yet the arid zone is still faced by other problems: locusts, for instance, used to invade the pasturelands until 1934. The warfare against this pest is going on by locating the hatching places and spraying from planes. The introduction of new plants, too, has not always been successful; for instance, the cactus, which was introduced as an ornamental and a fencing plant, spread in 1935 over 1·6 million ha (4 million acres) of pastureland, as well as over cultivated fields in the Cape country. Only by biological warfare, using parasites which destroy the plant, has this menace been exterminated.[1]

The northern margins of the desert belt: The semi-arid zone in the Magreb

Northern Africa (the Magreb), which includes Morocco and parts of Algeria and Tunisia, is situated on the Sahara's margins; but much of it is only semi-arid and not desert (Fig. 11.5). Most of the Magreb depends for its agriculture and water sources on rainfall only, and is a characteristic transition area—from the climatic, economic, ethnic and historical points of view—between the desert and the Mediterranean regions. The semi-arid zone occupies four-fifths of the area of North Africa (Despois, 1961). Within this area, however, there are more humid regions in the north, with 400–600 mm (15·5–23·5 in) of rain, which can almost be regarded as

---

[1] The cactus was *Opuntia megacantha*. Talbot (1961, p. 326) says, 'it was branded as the most noxious weed in South Africa; it occupied in 1935 two million acres, causing death to thousands of animals and had become a national menace'. Yet in the dry area of northeastern Brazil the 'prickly pear' is actually grown as cattle fodder! (Editor's note).

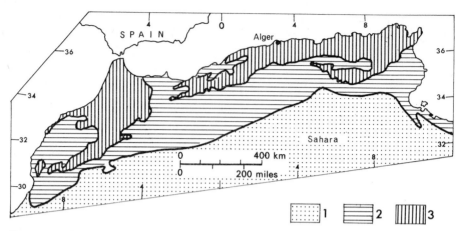

Fig. 11.5. The arid zone in northwest Africa (Despois, 1961).
1. Arid. 2. Semi-arid. 3. Mediterranean climate.

belonging to the Mediterranean zone, and southern regions with only 200–400 mm (7·8–15·5 in) of rain. The principal difference between those two regions is not so much in the amounts of rain, as in the reliability of these amounts and their availability (Fig. 1.2). The flora is Irano-Turanian steppe vegetation; this is a pioneer region, the domain of the desert's advance or retreat, according to the methods of land cultivation and the conservation of soil and vegetation. This is also the area where sedentary and nomadic societies and modes of life meet. Here, where the summer is prolonged and dry, and the winter brings only 200–300 mm (7·8–11·8 in) of rain, irrigation is compulsory for any crop. Annual requirements are: wheat 2 000–4 000 cu m per ha (28 500–57 000 cu ft per acre); fruit trees, 600–800 cu m per ha (8 600–11 440 cu ft per acre) a year; and vegetables 1 000–1 500 cu m per ha (14 250–21 325 cu ft per acre). This consumption is only a half or even a third of what these crops require in the desert zone proper.

Although the discovery of oil and phosphates forms an important and developing element in the Magreb's economy, the basic occupation of man in the semi-arid zone has remained rural—agriculture or stock-breeding. The French colonial settlement, which served at the time as an important economic stimulus for agriculture, had its hold mainly in the Mediterranean zone; the traditional modes of life still prevail in the semi-arid zone of North Africa, as they have done for many centuries (Despois Raynal, 1967).

The traditional land utilisation

Rational land use in the semi-arid zone, on the margins of the Mediter-ranean area, is possible only by employing balanced methods of dry farming, by using artificial irrigation from springs, as well as from wells and flood water, and by soil conservation on slopes through terracing. These methods have not changed noticeably in hundreds, possibly thousands, of years.

The greater part of the semi-arid land in the Magreb is under unirrigated cultivation, including not only wheat and barley, but also trees, such as the olive and fig. The cereals are grown in a biennial cycle, with intervals of fallow when the land is used for grazing. This is an extensive agricultural economy, which involves cereal growing, as well as the breeding of cattle, sheep and goats. In the traditional system the farmer waits for the first rain to soften up the soil; then he sows and covers the seeds by ploughing over them. Some do superficial ploughing before sowing; there is an advantage in superficial ploughing, particularly on slopes, in that it does not cause soil erosion.

The amount of seed put into the ground by the farmer is in reverse proportion to the aridity of the region: while in the north of the region 100 kg of seeds are sown on every ha (88 lb/acre), farther to the south only 80 kg per ha (71 lb per acre), and in the sandy southern area only 3 kg per ha (2·7 lb per acre); thus the farmer, who is sometimes more of a herdsman, may fortify himself against complete disaster in case of drought. Barley is preferred in the more arid regions, as it consumes less water and ripens three weeks earlier than wheat, thus escaping the dry hot winds which penetrate the area from the desert toward the end of the humid season; the barley can be reaped as early as the end of April or the beginning of May. Reaping in all areas is done with a sickle.

The borderline between the semi-arid and the Mediterranean zones is approximately the 400 mm (15·5 in) isohyet; that is the limit beyond which the growing of cereals becomes unprofitable. This does not mean that cereals and other crops are not grown also on the arid side of this isohyet; in the semi-arid zone, between the isohyet of 400 mm (15·5 in) and 150 mm (6 in), cereals are still grown, but fallow is also maintained there; the economy is that of extensive pastoral agriculture dominated by the breeding of sheep and goats, with no cattle. Beyond this area starts the arid zone, where pastoral or agricultural land use is impossible without artificial water supply and irrigation.

Traditional modes of subsistence and methods of cultivation exist side by side with modern methods which are the products of the French colonization. Man has lived in this area for a very long time: there is evidence of agricultural subsistence from the third millennium BC, but evidences of rock-painting show that pastoral subsistence existed even in the Neolithic Age. Ploughing is still mostly done with a wooden plough equipped with an iron wedge and drawn by various draught animals from the camel to the donkey. Between sowing and reaping the fields are neglected.

The breeding of sheep and goats is actually the only way to use the uncultivated soils in the steppe area, and it seems that methods have not changed since the Neolithic age. There is no definite nomadism here; yet there is transhumance, that is, regular migration along fixed routes towards fixed destinations, motivated by the winter cold and the summer drought. This movement is not only vertical, from the mountains to the plains, but also lateral, from certain water sources to others. The sheep and goats, or the camels, are the herdsmen's only property; they do not possess anything that cannot be carried. Pastoral life does not exclude land cultivation. On the margins of the steppe certain crops are grown; the necessity to go back and stay for a comparatively long period at one place, makes people semi-sedentary.

This pastoral-agricultural economy was much influenced by European settlers. In 1954, when agricultural expansion was at its climax, the French farmers in the entire Magreb possessed 34 000 farms extending over 5·4 million ha (13·3 million acres). Ten years later all these farms were nationalised or confiscated. The majority of the farms were in the Magreb's Mediterranean zone yet it is clear that the marginal region was also greatly influenced by the advanced economy, with its agrotechnical expertise, cultivation methods, new species, etc. For instance, wheat of the soft variety was introduced; the practice of ploughing before sowing, which was unknown to the native agriculture, started to spread toward the end of the nineteenth century. The construction of concrete channels and dams provided the basis for various irrigation projects and the control of springs; there was a considerable progress in well-drilling, particularly in the artesian basins. On the other hand, developments in sheep and goat breeding remained very small. There was actually some improvement through the selection of certain breeds, but no great advances were made in the protection of the pastures from overgrazing; the main directions of progress were in the provision of new waterpoints and the building of troughs, and in the combatting of pests and animal diseases.

During the 'colonial' period, then, two economic standards developed here: that of the French settlers, whose economy was based on export (including Algerian wines), and that of the native agriculture, based on crops for consumption such as cereals, figs and dates. On the one hand, an economy based on capital, and on the other hand an economy paralysed by absence of capital. On the one hand, large fields and farms, arranged geometrically, on the other hand, crowded villages and small irregular fields.

Since the termination of the colonial regime, the three Magreb countries have started to reorganise rural life and have launched various economic and social reforms. The fastest and most rigid reform has been effected in Algeria. The policy was to establish self-administration of the property abandoned by the original settlers; the agricultural labourers on the farms were called upon to manage them by their own councils under government control. All this refers more to the Magreb's Mediterranean zone; in the semi-arid zone, cooperative units were formed and each received 2 000 to 4 000 ha (5 000 to 10 000 acre) in which they grew mainly olives without irrigation, and bred livestock. Nevertheless a decisive change, such as delivering water from a distance, or deep drilling projects that could have turned the pastoral economy of the steppe into an intensive agricultural economy, has not yet reached the Magreb's semi-arid zone (Despois, 1967).

The Bedouin borderland in the Judaea desert: integration with the existing agricultural margins

The settlement of nomadic shepherds in the central Judaea desert, east of Bethlehem, is a good example of the mutual influences between rural agricultural settlements and the adjacent Bedouin area without a topographical or political barrier between them. Mutual influences of farmer and nomad, a transition from one way of life to another, while maintaining the ethnic duality, have existed here for thousands of years; but the radical change, that is, the transition from tents to stone built houses, and the formation of new villages which are dependent on pasture, as well as on marginal agriculture, occurred only during the 1960s.

The Bedouin domain on the border of the Judaea mountain massif (Fig. 11.6) in the Bethlehem area, extends over some 50 000 ha (124 000 acres). Fifteen per cent of this area lies in the Mediterranean phytogeographic region, 25 per cent in the Irano-Turanian (i.e. graminacea); the remaining 60 per cent are in the Saharo-Sindian or desert region (Shmueli, 1970). These pasturelands can bear, in accordance with their nutritive value, about 10 000 sheep or 19 000 goats. Some 13 200 tribesmen, constituting about 25 per cent of the total population of the Bethlehem area, were living in this area in 1967. The remaining 75 per cent comprise the farmers and the urban population of Bethlehem.

A study of the tribesmen's settlement process will show that one of the elements of the initial settlement was the division of landed property. In the eighteenth and nineteenth centuries, according to contemporary testimonies of travellers, the tribes of this region were already cultivating those parts of the borderlands which were suitable for cultivation; this practice is characteristic of semi-nomadic tribes, such as those under discussion. The short range of migration in the steppe region, which is very narrow in Judaea, was also a factor in regulating the migration habits. Since land cultivation became permanent, it has become necessary to settle the problems of land ownership. Demographic pressure caused the enlargement of cultivated areas by annexation of pasturelands. Until the beginning of the British Mandate in 1920 the matter of ownership was ordinarily settled by the tribe itself with no governmental interference. Three types of ownership prevailed. Private lands, which were first divided at the end of the eighteenth century or the beginning of the nineteenth; lands purchased by the tribesmen later from sedentary inhabitants; and communal

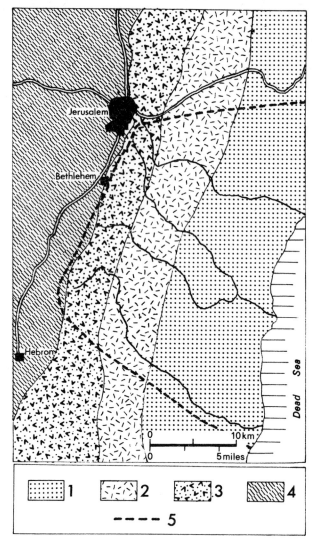

FIG. 11.6. Phytogeographic regions in Judaea (Shmueli, 1970).
1. Saharo-Sindian desert region. 2. Irano-Turanian steppes. 3. Mediterranean garrigue.
4. Mediterranean scrub. 5. The area of Bedouin tribes considered in this study.

land belonging to the tribe as a whole. The final division of lands, including
the tribal lands, was done during the early 1930s, accompanying the land
registration carried out by the authorities.

The settlement of the nomads in the Bethlehem area was spontaneous
and without intervention from external elements. This was not accomplish-
ed by a single act; it has been a process which has been going on since the
1920s and is not yet complete. The influence of contact with the permanent agricultural and urban settlements might have spurred this process

on, but several factors have probably been involved (Shmueli, 1970).

The central government took a firmer hold toward the end of the Ottoman period, and even more so during the British Mandate. The government even assisted the Bedouin with food and fodder, in times of drought. Another factor of long standing has been the bond between the nomad and his neighbour, the falah. These were contacts of employment, trade and services. With the country's development a flow of surplus labour from the tribes was directed to the villages, and even to the cities. The main reasons for permanent settlement seem to have been preference for permanent dwellings as opposed to tents, weariness of nomadic life, and the desire to give the children proper education. The transition to sedentary life took several forms.

Hired labour in agriculture in exchange for a quarter of the harvest

The tribesmen used to go to Trans-Jordan for seasonal work in agriculture. They did any sort of work—ploughing, sowing, reaping and threshing— and received in return a quarter of the harvest. Every tribe from the area in question sent a number of men to do these jobs. There were also reverse cases, when tribesmen gave their lands to their neighbours, the farmers, for cultivation in exchange for a quarter of the harvest. These seasonal jobs came to an end in the early 1930s, since employment could be found on the spot.

Transport by caravan

This was a characteristic occupation in the late nineteenth century and the first two decades of the present century. The crops from the Jordan's eastern bank were in demand on the western bank and since at that time motor transport was not developed, the country was not divided, and the population on the western bank needed wheat, barley, beans and other crops from the eastern bank, the Bedouin undertook the organising of caravans. These caravans, of twenty or thirty men each, were organised on a tribal basis. One round trip from Bethlehem to Amman took about eight days, including the purchase and gathering of the crops on the spot and the return trip to the markets of Bethlehem or Jerusalem. These caravans disappeared in the late 1920s because of the political severance of the eastern and western banks, and also because of the increase of motorised transport. The gathering of salt on the shores of the Dead Sea employed some tens of workers during the 'dead' seasons in agriculture, or in the breeding of sheep and goats. The salt was sold in the cities of Judaea and also in Trans-Jordan. This occupation persisted until the end of the British Mandate. Jobs in building and in road construction, which were associated with the development of Jerusalem, attracted labour from among the tribesmen, from the early twenties. The principal jobs were in quarries and in the transport of building materials. The increase in building for the British army during the Second World War continued and enlarged this field of occupation.

The process of settlement

The process is well on the way to completion; the exchange of the tent for a permanent dwelling is its major feature. The first house was built in the 1920s and the rate of building was slow; between 1954 and 1958 the pace grew quicker. The transition from the tent to a permanent dwelling was a revolutionary step for the Bedouin; yet it appears that the influence of the tent is still noticeable in the shape of the village houses. Many houses are built so that their walls are only a substitute for the tent canvas, with the door facing east, just like the opening of the traditional tent. The dwelling habits inside the house are identical to those within the tent.

The settlement process has been expressed by the decreasing importance of sheep and goat breeding, and the increase of cultivation. Thus the number of sheep in the tribal areas, that towards the end of the British Mandate had reached 62 500 declined to 28 000 during the Jordanean rule; the same happened in the case of goats which decreased from 26 000 to 13 000. On the other hand, the cultivated area grew in the tribal area which is located, as mentioned before, in the Irano-Turanian phyto-geographic region, and does not use artificial irrigation. The difference between 1945 and 1967 was overwhelming: the unirrigated field crops increased by 235 per cent (from 380 to 937 ha/940 to 2 320 acres). Vegetable gardens increased from zero to 25 ha (60 acres), orchards increased by 324 per cent from 138 to 586 ha (340 to 1 440 acres). These figures are the most significant illustrations of the radical change in the way of life.

Summary

It is apparent that the main incentive for the socio-economic changes in the Bedouin borderland in the Judaea desert, was the proximity of sedentary settlements, and the facility of adaptation. Moreover, the element of outside employment, which brought in money, provided the capital for the investments that were needed for purchasing lands, tools, seeds and building material. If there had not been a contact with, and influence from, the world outside, mainly through trading, the borderland would have remained in its former backwardness.

12
Basins at the foot of well-watered mountains

The environment discussed in this section enjoys a distinct advantage over the regions where direct rainfall is the only source of water. It lies near to regions with abundant precipitation and favourable relief and structural conditions for water supply. These resources are, in most cases, distant or too deep for immediate tapping. There is therefore a great differentiation in the degree of their exploitation as between physically similar regions. In many regions there is not as yet a consciousness of the existence of these resources; even where it does exist, there is very little application to usage. Technology is not the determining factor in this case; it is history and culture that matter.

Pirzada: an Afghan village feeding on qanat water

The Pirzada oasis is situated 65 km (40 miles) west of Kandahar (Humlum, 1959), on the border between the steppe and the desert (Fig. 12.2). It is 930 m (3 000 ft) above sea level, on the bank of the Kutchinkud stream, whose source is in the nearby Sakh Maksud mountains. Winter temperatures do not fall below zero, the average being 7°C (44°F), whereas the summer is very hot, the average temperature being 35°C (95°F). Rain falls in winter, with the arrival of the most easterly Mediterranean depressions, and reaches 150–200 mm (6–7·8 in) in a season. The centre of the oasis is an ancient emporium (caravanserai), and a market on the spot where the road crosses the stream channel; these two institutions are declining.

Methods of water supply

The only source of water supply are the qanats, which start at a distance of about 5–10 km (3–6 miles), at the foot of the mountains, where they reach groundwater at a depth of 20–50 m (65–165 ft) below the surface. From this starting point, the water flows, in underground conduits, until it reappears on the surface, at the centre of the oasis, and there runs in open channels. (Fig. 8.1). This oasis is not an isolated phenomenon in the area, but one of a large group of oases subsisting on the same method of water supply.

The stream which flows near the village is not reliable; although its channel breadth in the Pirzada are reached 250 m (273 yd), the flow in it is sporadic; it occurs rarely and persists for a few hours only—a characteristic flow for this hydroclimatic zone. There is no flow whatsoever between March and November, when the stream bed is completely dry. The farmers' water rights differ according to their wealth; some are entitled to twenty-four or even thirty-six hours of irrigation per week, whereas others are entitled to only one or two hours a week (cf. p. 121). Some qanats start at the stream banks in the alluvia in which the groundwater table is found at depth.

The fields

The plots are arranged in accordance with the surface gradient: they are long (50–200 m : 55–220 yd) and narrow (a few metres). They lie parallel to the qanats and the channels run in a north to south direction. The water in the qanats and the ditches flows sluggishly, and the inundated field is actually an extension of the ditch. The irrigation ditches (as distinguished from conduits) are of an interdigitated pattern, and it is sometimes difficult to determine whether the area is a field or a channel (Fig. 12.1). There are

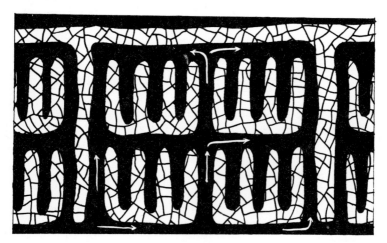

FIG. 12.1.   Cotton irrigation methods in Afghanistan's oases (Humlum, 1959). The arrows indicate the direction of flow of the water.

quite a number of land ownership patterns. The most outstanding differences are in the size of land property; there is also a considerable number of tenants who till the land in exchange for a part of the harvest, ranging from one-tenth to one-half with an average of one-quarter. Here, where the treatment of land and water is so delicate, there are very ancient rules concerning the use of water.

Methods of cultivation

Methods of cultivation are tranditional and there have been no improvements of either tools or seeds; the poor yields, ranging from 500 to 1 500 kg (450 to 1 350 lb per acre) of wheat or barley per hectare, are due to that handicap, as well as to water shortage. Fertilisers are not used, ploughing is superficial, turning only a layer of 10 cm (4 in); the seeds are scattered by hand from above and covered with earth by a plough. Reaping is done with a sickle, threshing by oxen, and the grains are separated from the straw with a sieve. Generally, two crops are possible: in winter wheat and barley are grown, cotton and sesame in summer. The harvest seasons are the spring and the autumn.

Winter crops

Wheat and barley are sown in the autumn, usually in October, and harvested in May–June. Although rains fall during the growing season, they do not suffice, and it is necessary to irrigate several times during the winter, about every fifteen days.

Summer crops

The summer heat makes it necessary to determine preferences for claims on water. In the first place come orchards, vineyards, melons and various vegetables; then come maize and cotton; rice, which is a characteristic crop of south-eastern Asia, is not grown in oases which feed on the qanat water since the amount of water is too limited here to allow a crop with such a high demand.

Irrigation methods

The area of irrigated lands in Afghanistan today is apparently smaller than it was before the thirteenth century; part of the ancient systems were not reconstructed after the Mongol occupation. Information concerning the size of the irrigated area varies widely—estimates running from 20 000 to 500 000 ha (50 000 to 125 000 acres) (Fig. 12.2).

Since the groundwater table exists at varying depths, mostly several metres and even tens of metres, water is seldom drawn up by the shaduf method, which is so common in neighbouring Pakistan. Irrigation from rivers exists only in north Afghanistan, in the high mountains and the semi-humid zone, or in the desert area in the south of the country from the water of the river Helmand. Everywhere else in the country irrigation is based entirely on the qanats. The qanat culture in Afghanistan is very ancient and is still highly developed; in contrast to many other regions where the qanats are being abandoned, in Afghanistan they are carefully

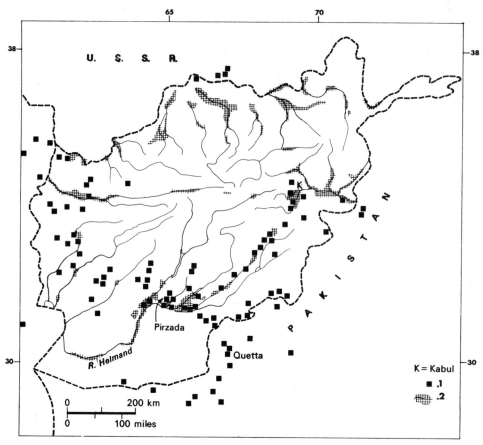

FIG. 12.2.   Irrigation enterprises in Afghanistan (Humlum, 1959).
1. Localities irrigated by qanats. 2. Localities irrigated by streams.

maintained. It is assumed that some have been in use for 2 000 to 3 000 years; in many areas it is only the existence of the qanats that makes permanent human occupation possible. Most of them are in the southern part of the country; north of the Hindu Kush range the easier method of irrigation from streams is possible, and there are few qanats.

Irrigation is performed in small limited plots, enclosed by small dirt dams; the width of the plots is only 3 to 4 metres (10 to 13 ft), but the length sometimes reaches a few hundred metres. Since the gradient is quite steep (usually, 1 per cent, as most of the lands are in inland basins and pediments) —it is braked by the bends in the ditches. This method—named 'pulwan'— is applied to the irrigation of wheat and barley; maize and cotton are irrigated in small basins where the water stands for some time (cf. Fig. 12.1); this system has a configuration of a 'comb' (Humlum, 1959).

The system of water distribution among the inhabitants

There is no life without water in the oases; there are therefore fixed rules laid down by the Koran for the use of water, and it is to be assumed that the Koran drew these rules from more ancient sources. River water is regarded as a gift from Heaven, and as such cannot belong to one man; qanat water—obtained as it is by special efforts from sources which were not previously available—belongs to the man, or group of people, who dug the qanat, so that qanat ownership, private or collective (of a family or a village), is clearly defined. Water stealing is considered a most grave offence.

The division of irrigation time in Pirzada may be illustrated by the following example: eight people own the Pirzada qanat; in the course of eight days each of the owners has the right to use once all the water of the qanat. But since the rights are not divided equally among the eight owners, some have only the use of part of a day in the course of the eight days round, whereas others can draw water for more than a day during the same round.

This sort of water division does not take into account the different needs of crops which must be irrigated more frequently and those that need less frequent watering. As will presently be shown by the example of Beth-She'an in Mandate times (cf. p. 142), this system of arbitrary distribution of water according to the share in ownership, instead of adapting the irrigation system to the needs of the crops, hinders changes in cultivation methods and the introduction of new crops.

Summary

The Pirzada oasis, as a sample of life in a closed basin which feeds on the qanat water, is shown up as a settlement where traditional methods have been perfected to the maximum. Moreover, as long as the country is generally agricultural in nature and does not contain a population with a high level of consumption it is to be doubted whether the situation in Pirzada will change, and whether there will be intensification of crops. The mere presence of water resources with the ancient methods for employing them, is not sufficient to effect a decisive turn towards modern economy. Only the diversion of the economy toward commercial cropping can make better use of the region's potential.

The Fergana valley: an intermontane basin on the Syr Darya

The Syr Darya river is one of the most important rivers in the USSR irrigation system. Its water irrigates about 3·2 million ha (8 million acres) of lands, mostly dedicated to growing cotton; this is approximately a third of the total irrigated land in the USSR—which in 1965 was about 10 million ha (24·7 million acres). More use is made of water of the Syr Darya than of the neighbouring Amu Darya, as there are no available soils suitable for agriculture in the latter basin. The Syr Darya valley, on the other hand, contains soils which are suitable for agricultural cultivation, particularly alluvial fans and loess deposits. Irrigation is maintained by ditches, some of them very ancient, particularly in the upper reaches of the river, whereas modern irrigation ditches have been constructed down river and development is spreading from the Fergana valley toward the 'famine steppe'.

The natural conditions and ancient form of irrigation

The Fergana valley is an elliptical-shaped intermontane basin (Carrière, 1966), 300 km (186 miles) in length and some 100 km (60 miles) in width (Fig. 12.3). The valley is surrounded by mountains, over 5 000 m (16 400 ft) high, and protected against the invasion of particularly cold or hot air flows. The valley bottom is 500 m (1 640 ft) above sea level in the east, and 300 m (985 ft) in the west; most of it is composed of an agglomeration of detrital deposits which were laid down in the basin during the Tertiary and the Quaternary periods. The general physiography comprises river terraces in the centre and alluvial fans at the edges. The bottom is overlaid with Quaternary loess deposits a few metres thick, which yield fertile soils. There is a distinct difference, however, between the well drained soils on the margins and the soils in the centre which are less well drained and which tend to develop crusts by concentrations of calcium and magnesium. Where the drainage is bad, soils of the solonetz and solonchak type appear. These can be put to agricultural use only after being leached of minerals and by lowering the water table.

The region is in the semi-arid zone, with a cold winter; though the Fergana valley is protected against invasions of cold polar air masses toward the end of the cold season, inversions are liable to produce quite low temperatures. During the transitional seasons between the hot summer and the cold winter, rain bearing depressions pass through the region. The amount of precipitation does not exceed 300 mm (11·8 in), 50 mm (2 in) of which fall during the hot season. Agricultural crops can be grown only with irrigation from river water.

The streams that cross the Fergana valley receive a regular supply of water from melted snow. Their regime coincides with the demands of agriculture, as their highest discharge occurs in spring, when water consumption is at its highest; in summer the discharge is abundant enough for maintaining the crops during the entire growing period. Irrigation ditches apparently existed here as early as the fifth millennium BC. Civilisations based on irrigation flourished during many periods in history, particularly in the fourteenth and fifteenth centuries AD. The annexation of the valley by Russia in 1866 marked the renewal of the irrigation activity, especially in the cotton fields which were supplying the textile factories of

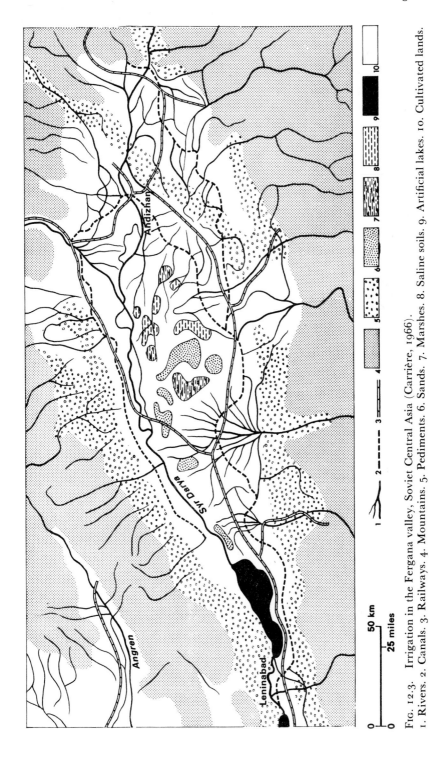

FIG. 12.3. Irrigation in the Fergana valley, Soviet Central Asia (Carrière, 1966).
1. Rivers. 2. Canals. 3. Railways. 4. Mountains. 5. Pediments. 6. Sands. 7. Marshes. 8. Saline soils. 9. Artificial lakes. 10. Cultivated lands.

Tashkent is just off the northwest corner of this map. A – Andizhan, L – Leninabad.

Russia during the Civil War in America, when the cotton export from the US South was cut off. In 1897 a railway was constructed to the Fergana valley, and in 1910 Fergana produced 70 per cent of Russia's total cotton crop.

The possibilities of irrigation in the 'famine steppe', along the course of the Syr Darya but outside the boundaries of the Fergana valley, have not eluded the authorities, and in 1879 a new canal, 12 km (7 miles) long, was brought into use—the Kanpemensk canal; another canal, 25 km (15·5 miles) long, followed in 1891, and in 1900 the Serdevsk canal, 84 km (52 miles) long. Most of the lands which were thus made available for agriculture were divided among Russian settlers. In 1913 the irrigated area in the 'famine steppe' reached 50 000 ha (123 550 acres).

After the 1917 Revolution the efforts to increase the economic strength of the region were continued. The first years were dedicated to reorganising the economic system and to agricultural reform. The reforms were applied not only to the division of lands, but also to the division of water rights; in Uzbekistan, for instance, about 200 000 ha (495 000 acres) of irrigated land were divided among 50 000 families.

An expansion of the irrigation system occurred in 1939, when a large labour force was mobilised, in a short time, to build 500 km (310 miles) length of irrigation ditches, particularly in the Fergana valley, where an effort was made to unite the ditches that ran down the alluvial fans into one central channel. The result was the 'Great Canal of Fergana', 270 km (168 miles) long, that started in the Narin river and terminated in the southern part of the valley. All the dams are built of earth. The construction of the canal enabled increased irrigation water to be applied to an area of 50 000 ha (123 500 acres), whilst the total irrigated area was enlarged by 60 000 ha (148 250 acres). The irrigated area belongs to three republics —the Uzbek, the Tajik, and the Kirgiz republics, and is managed jointly by all three.

Other channels are the northern canal which also derives from the Narin and irrigates an area of 65 000 ha (160 600 acres), and the southern canal which irrigates a somewhat larger area of 70 000 ha (173 000 acres). These two canals were constructed in 1950 (Lewis, 1962).

Water from the Syr Darya

The Second World War slackened the pace of development in the hydraulic installations. After the war, the nature of the development work underwent a change: instead of enterprises which had been executed by a single effort involving a large expenditure of manpower, the Syr Darya was now regulated by a series of modern dams that could be used for creating energy for industrial plants as well as for irrigation. This big effort was undertaken jointly by the three republics concerned and aided by an allocation out of the general budget of the USSR. From the Farkad Dam a canal with a capacity of 470 cu m (550 cu yd)/sec, goes into the Dalverzin desert. The dam holds about 250 million cu m (283 million cu yd) and creates a lake 41 sq km (25·5 sq miles) in area. But the volume of water is decreasing as a result of silting, and if the present rate persists, the lake is expected to have a life of only thirty years; the river discharge is 17 000 million cu m (22 000 million cu yd) a year; but this amount includes 32 million tonnes of silt of which 50 to 90 per cent accumulates on the bottom of the artificial lake.

Another dam is located near Leninabad in the Tajik Republic, at the spot where the valley narrows and the Syr Darya emerges from it. Here, a dam 1 000 m (1 094 yd) long was built behind which is a lake of 500 sq km (310 sq miles) holding 4 000 million cu m (5 000 million cu yd) of water. The river's discharge is 16 000 million cu m (20 000 million cu yd) a year, which allows the irrigation of 220 000 ha (544 000 acres), out of which 100 000 ha (247 000 acres) are the virgin lands of the 'famine steppe'. The silting up of the reservoir is expected in sixty years.

The organisation of irrigation

There is a special ministry which deals with the improvement of soils and irrigation. There is no irrigation by sprinkling; most of the irrigation is done in furrows. Agricultural towns are being planned—each possessing 8 000–10 000 ha (20 000–25 000 acres), which are designated for growing cotton, and orchards. The Leninabad *sovchoz* (near the town of this name) is an example: in 1966 they cultivated 8 000 ha (20 000 acres) of which 6 000 ha (15 000 acres) were cotton fields, 500 ha (1 235 acres) alfalfa, 300 ha (740 acres) vegetables and fruits. This production was effected by 1 100 workers, using 164 cotton-picking machines and 420 tractors.

Tashkent, on the Syr Darya, at a distance of 150 km (95 miles) from the river's outlet from the Fergana valley, serves as a good example of a city in the semi-arid zone that subsists entirely on river water (George, 1956). It came into existence through the utilization of the waters of two rivers, the Chirchik and the Angren, at the point where they emerge from the Tien Shan mountains which rise to over 4 000 m (13 100 ft).

Until recently the oasis appeared as a fan of irrigation ditches. The work which had been done there during the last fifty years, and particularly the last thirty years, changed the oasis fundamentally, through the use of the Syr Darya water in the 'famine steppe' area. By 1940, 250 000 ha (over 620 000 acres) within the oasis were irrigated. Since then the area has been doubled and now amounts to 500 000 ha (nearly 1·25 million acres). Sixty per cent of this area is occupied by cotton fields, the remainder by rice, and mulberry trees for silk worms. Wheat and barley are grown on the oasis margins, without irrigation.

The change that has come over the city is not only in the increased amount of water supply and the enlargement of the cultivated area, but in its economic character. Until the arrival of the Russians in 1885, this had actually been a crowded orchard village, and not a town with urban functions. It had 200 000 inhabitants, the houses were built of mud and were closely packed. On its margins a new city was built; 1 000 Russians lived there toward the end of the century, on an area much bigger than that of the Uzbek town. The main irrigation canal separated the two cities.

In 1925 the two parts were united with the intention of forming an ethnic and communal assimilation. In 1939 there were over half a million citizens in the town; it was still very scattered in form, but its urban functions had increased. Tashkent has since become the biggest city in central Asia—a centre of air routes, railways, food industries and cotton ginning. In its neighbourhood power stations have been built on the rivers that come down from the mountains and also textile mills and an agricultural machinery industry. Today half the active population is employed in industry and the population is approaching 2 millions.

Summary

The Fergana valley, the 'famine steppe' and the city of Tashkent, are not isolated phenomena in the semi-arid zone; they are a part of a great country with immense economic, industrial and technological resources. Their development is due not only to the available water from the surrounding mountains, but also to the ability to turn this water into an economic lever, in this case, for growing cotton for the whole of the USSR. The entire structure of agriculture has changed thereby: instead of subsistence agriculture and auxiliary irrigation, new crops have been introduced, as well as mechanisation on a large scale, agrotechnical expertise, and last but not least, an industry which absorbs the agricultural products. Water utilisation has been only a part of the region's economic revolution.

The challenges of the semi-arid environment in southern California

The desert and semidesert areas in the southwest of the United States of America in general, and in California in particular, provide an almost unparalleled opportunity to study the development of an area of this type towards an advanced urban and technological civilisation (Hodge, 1963). Most of the region is semi-arid; extreme arid areas are few.

Various civilisations have been active in the arid southwest: the Indian civilisations were concerned primarily with ensuring mere existence. The first agricultural civilisations, in the first centuries BC, developed dry farming on the Colorado plateaus together with an irrigation civilisation which collected runoff from the slopes.

The Spanish carried on their Mediterranean type of irrigated agriculture in the big river valleys. The Spanish, for whom the present southwest of the United States was the most distant area from their centre of activity, brought with them sheep and goats and horses, but did not engage in large-scale colonisation. Towards the end of the Spanish rule, only a few settlements, mainly mission stations, existed.

Stages in the agricultural development of southern California before 1940

The enormous development of Californian agriculture, which until the Second World War was the main contributor to the State's prosperity, cannot be grasped without a study of the various stages in the development of agricultural technology, in water supply, as well as in the introduction and acclimatisation of new species and crops, the development of industrial technology in conservation, processing and transport of agricultural products. The first irrigation project using brush and earth dams and tile pipes, was carried out by the Spanish in 1769 (Raup, 1959). The first artesian well was dug about a hundred years later, and a water-pump operated by a steam engine, which could draw water from considerable depths, was put to work in 1872. The use of concrete and steel pipes soon followed, and water pumping by electric engines was first started in 1885. Next came the period of dams: the Bear Valley project in 1884, the Yuma project in 1904, and the All-American Canal in 1940. These projects

Plate 14. Trucks of alfalfa hay for urban dairy 'farms', Los Angeles (courtesy: S. H. Beaver).

brought to the region water and energy from the snow-covered mountains.

Long distance transport facilities were a prerequisite for the commercial-isation of agriculture. The three main railways between the west and the east of the continent were constructed in the years 1869–85; refrigerated wagons were put into use as early as 1880; fifty years later, in 1930, trans-portation by refrigerated road vehicles started. Along with this technology, which had no parallel in any other part of the world at that time, appeared methods for preserving the products against the hardships of climate— such as oil-fired orchard heaters. We must bear in mind, that nineteenth-century California was not urbanised as were the states of the Union. From the 1860s, agricultural products were preserved by artificial de-hydration, by the use of tin-cans, and by refrigeration, and in the 1930s the quick-freezing process was developed.

During the second part of the nineteenth century, a greater variety of agricultural crops was introduced; hitherto the crops were of the Mediter-ranean variety, brought by the Spanish in the 1760s: citrus, olives, vines, figs and wheat. In the latter half of the nineteenth century, alfalfa (for fodder) was brought in and acclimatised, and after that came cotton, almonds and the melon.

The period since the Second World War has been one of accelerated industrialisation in southern California, but despite the rapid urbanisation which absorbs agricultural lands, the agricultural landscape has not been obliterated. The farmers of southern California have the advantage of a highly developed air transport system, as well as highways and a large fleet of refrigerator trucks to carry delicate agricultural products. In a

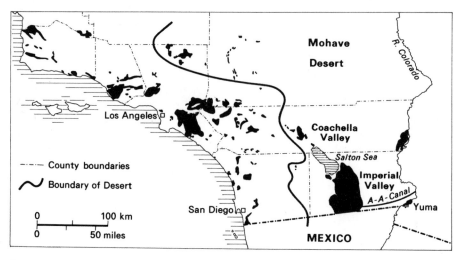

Fig. 12.4. Irrigated lands in southern California (Raup, 1959).
(Aschmann, in Thomas, 1959)

very short time the product passes from the production area to its destination in the east.

This development—excluding the first projects, such as the Yuma project, or the All-American Canal—has been supported by local and not Federal funds. The water for agriculture is supplied by drilling, as well as by the damming of inconspicuous small rivers. Today California is first in the United States for fruit production and among the first for milk, sheep and goats, and egg production. The areas of dry farming, however, are not large and occupy less than 10 per cent of the total area. A similar area is occupied by extensive pasture (Fig. 12.4).

In the general development of agriculture there is a certain regression in citrus growing, which in the 1960s only represented 6 per cent of the total value of agricultural production; citrus groves were the first to be turned into urban zones. The urbanization in the Los Angeles district has indeed reduced the agricultural area, but it has not reduced the economic value of agriculture. A thorough change has occurred here, with a decrease of citrus orchards on the one hand, and an increase of dairy and poultry production, on the other—an intensification of agriculture in response to the demands of the large urban population.[1]

The change that has occurred in California during the last hundred years, turning a landscape of pastureland into irrigated fields, has been most advantageous. It has been effected by a combination of several factors—technology, urbanisation and transportation; and the growth of local markets in the state itself, supplementing the distant markets in the east, has contributed appreciable to this change.

Urban development

Agricultural development went side by side with one of the world's greatest urban developments, and there is a mutual reaction between the two. The twentieth century may one day be labelled the 'age of urbanisation', so different is the appearance of the world towards the end of the century from what it was at its beginning, and California is perhaps, the most striking example of this change. This is interesting, because in the beginning of emigration to California, its urban centres had no attraction for the emigrants; its special quality, which is favourable for urban and industrial development, has been recognised only in the last few decades (Nelson, 1959). Apparently, this development has had to await the advanced stage of technology to which the USA was ascending during and after the Second World War.

---

[1] Much of the milk consumed in Los Angeles comes from 'farms' within the city; these are little more than stock-yards for the cows to exercise in, plus a huge open barn for the storage of alfalfa hay imported in large trucks from the irrigated hinterland (see Plates 14 and 15). (Editor's note).

Plate 15. An urban dairy 'farm' in Los Angeles (courtesy: S. H. Beaver).

During that period population migration to southern California was one of the greatest movement in the history of the United States; most of the new inhabitants settled in towns. The rate of growth in the years 1950–60 was, in the Los Angeles Metropolitan area, about 200 000 a year, which constituted 35 per cent of the general growth in population in the eight years between 1950 and 1958 (not taking into account the natural growth). San Diego grew during the same period by 63 per cent. This is the highest rate of growth for the whole of the United States. The number of industrial employees grew by the same proportions: one-sixth of the total growth of employment in the United States, during the period is attributed to the Los Angeles area. As a result a complete change has come over the landscape of the coastal strip of southern California, the most remarkable semi-arid zone of the western United States.

As already noted, the area occupied by agriculture is consequently decreasing. The moment that a citrus grove is included in the metropolitan area the municipal taxes become so heavy that there is no sense in keeping it on as an agricultural land. Some crops, such as flowers and strawberries, were damaged by the heavy city smog, for which Los Angeles is famous (or infamous). The areas which are not cultivated, or, so to speak, unused, assume urban functions; forests become centres of recreation, the mountain slopes turn in winter into ski-courses, the streams are visited by amateur fishermen: it may be said that there is also an urbanisation of 'natural' areas.

The industrial development of southern California

The most prosperous period in the agricultural development of California was the second half of the nineteenth century, continuing until the end of the First World War. It is true, that the transition was not radical but gradual, until the early 1950s; until that time agriculture was the principal element in the economy of California; and though there has been a certain decline in some branches since then, the agricultural nature of the state has not been erased even by the accelerated urbanisation of the present day. However, the share of agricultural commodities in the total production of the state is being reduced. There has been several stages in this process of changing values. Some of the stages are not associated in any form with the semi-arid nature of the region (such as the discovery of oil), whereas other stages are definitely connected with the climatic conditions; the inter-relations of the various elements may be examined and conclusions drawn as to the particular role of the special climatic conditions. The oil boom period was one of the first to bring to California a flow of people and capital, when oil resources among the richest in the United States were discovered there. Although the presence of oil in the Los Angeles area had been known before, there was little production until 1917, when the development of the automobile industry increased the demand for fuel, and at the same time improved drilling equipment became available. The peak in the development of the Los Angeles area oil fields was reached between the years 1920 and 1924; this development resulted in no small measure from the fact that the nearest coal mine, the conventional source of energy, is 1 200 km (745 miles) away in Utah.

During the oil period two other industries came into existence, and in the course of time they have assumed a greater importance than oil. The development of both was attributable to a great extent to the climatic conditions. This may have been the first case in the world's history, in which climate formed a dominant factor not only for agriculture, as it has always been, but also for industry: the film industry and the aircraft industry. The film industry has profited here, not only by extensive empty and comparatively cheap areas, but also, and chiefly, by the large number of sunny days, which were at that time a prerequisite for filming. In addition, the young industry benefited by the greatest variety of landscapes at the shortest distances from the studios: beaches, desert, mountains and plains. Between 1913 and 1920 most film companies of these days moved over to southern California; the distance from markets is of no consequence to the film industry, as it is for heavy industries or agricultural production. The climatic advantages lost their importance when the techniques of photography were further developed; nevertheless, the concentration of studios and staffs continues to hold most of this industry, even today, although it is no longer as exclusive as it used to be. Though the number of employees does not exceed some dozens of thousands, the income from this industry is out of all proportion to the numbers engaged in it. The popularisation of California, which followed the film industry, has also augmented tourism and recreation in the area, and it was also a source of attraction for superior technical talents, who were prepared to exchange their places in the east of the United States for residences in California, when it was suggested by their employers in the electronics or aircraft industries.

The aircraft industry, which also needed free spaces, and suitable weather with little rain or snow, and with light breezes, particularly during the early stages of this industry, before the invention of sophisticated navigation instruments, was also attracted to southern California for obvious reasons. As early as 1928 California was already second only to New York in concentration of aircraft plants. But the main development came during the Second World War; after a certain decline with the cessation of hostilities, a renewed development started with the space projects during the 1960s. About 30 per cent of the employees in industry in the Los Angeles area during this decade were connected in one way or other with the aircraft industry; in San-Diego, the percentage rose to 75. Together the two cities contain a quarter of the entire United States aircraft industry. These industries require large spaces for test airfields and for storing machinery parts (Fig. 12.5). No wonder that in the mid-sixties, southern California received 5 thousand million dollars worth of orders a year for the military industry—more than the combined income from mines and agriculture. The increase in the average income per family was greater in arid California, in the years 1940–60, than in any other state in the Union (Fig. 12.6).

California as a residential area for the twenty-first century

If today California seems an attractive area for the twentieth century man, who likes sports and has free time to indulge in outdoor hobbies, who can move over long distances from home to work or to recreation areas, who can enjoy the fruits of his working life in a retirement on a pension, while residing in an area with a favourable climate, it is still true that the conversion of the natural environment into a comfortable residential area has required a great effort. The Mediterranean or semi-arid conditions presented man with the challenge of a new environment, different from those to which he had been accustomed in the earlier settled areas in the east of the continent; one of the first missions to be undertaken was the supply of water. Los Angeles receives water from Owens Lake, 500 km (310 miles) away; the Colorado aqueduct also delivers water from a distance of 300 km (185 miles). The relative proximity of the semi-arid region and the snow-covered mountains which provide a permanent source of water, is one of the reasons that keep the region as a developed inhabited area—while man with his technological devices and organising ability has made this development possible. In order to maintain a high standard of living in the semi-arid zone, large amounts of water are needed. Some

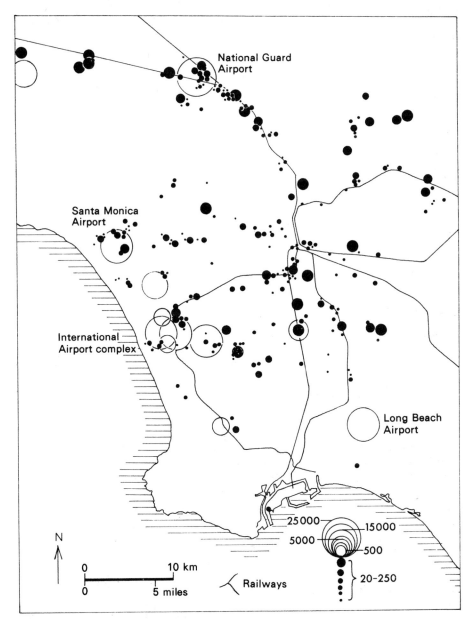

National Guard
Airport

Santa Monica
Airport

International
Airport complex

Long Beach
Airport

25000
15000
5000
500
20-250

N

0   10 km
0   5 miles

Railways

Fig. 12.5.   Industry in the Los Angeles area (Nelson, 1959). Symbols proportional to numbers employed: circles, aircraft manufacture; black discs, other industries.

cities require 1 cu m (1·3 cu yd) per person a day—but 40 per cent of this may be used to water lawns. Yet this water is usually not more expensive than water in humid regions (Wollman, 1962). There are also drawbacks

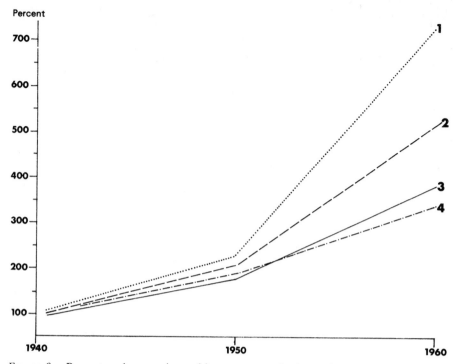

Fig. 12.6.   Percentage increase in total income per capita in southwestern USA, 1940–60 (Garsney & Wollman, 1963). 1. Arid California. 2. Mountain West. 3. United States. 4. Great Plains.

in this landscape: the Los Angeles river floods are to be expected, as in any river or stream in the semi-arid zone; the storm of 1938, which yielded 300–450 mm (11·8–17 in) of rain in five days, drowned eighty-seven people, and caused damage estimated at $78 million. During a flood, the discharge of the Los Angeles river, which has an intermittent flow, reaches volumes several times larger than the ordinary discharge of the Colorado. Such floods can of course be checked by dams, diversions and afforestation; soil conservation and flood control projects have considerably reduced the danger, though at very great expense. The central problem today, as regards the climate, is the photochemical smog, which finds here suitable atmospheric conditions, mostly in the form of the temperature inversion generated by the high atmospheric pressure in these latitudes. In Los Angeles the problem of environment has attained large proportions, because of its atmospheric conditions. Man solved the problem of aridity, but has not yet overcome the more general problem, that of the damage done to primary natural conditions by his economic activity. This problem is not exclusive to Los Angeles or the semi-arid zone; it is the problem of advanced urbanisation, all over the world.

13
Tapping of groundwater and springs

Although in most of the arid regions groundwater and springs—which are the valves of groundwater flow—are exploited in varying degrees, in some areas the modes of life have undergone a complete change as a result of the increased effort to make use of these sources. Naturally, great changes could be effected only in recent periods, as drawing up large amounts of water from deep groundwater tables is possible only with modern technology. Of all the ways to raise the value of the extreme arid zone, this seems to be the surest and most sophisticated if there is sufficient hydrological expertise and provided there is a sense of responsibility not only towards the present generation, but also towards future generations, so that measures are taken to preserve groundwater storages, and to prevent their disappearance through excessive extraction.

Artesian groundwater in Australia

Of all the semi-arid countries which are now in advanced technological stages, Australia was the first to advance on the arid zone with the tools and the aims of modern economy. It was the first to try artificial rain, and the early ripening species of wheat, suitable for the conditions of rapid growth during the short humid season, were developed there. This is not surprising, as more than three-quarters of the country lies on the arid side of the aridity boundary (Fig. A5, p. 173). The temperate zone is here quite narrow and limited in extent, in comparison with other countries on the border of aridity, such as the United States, the Soviet Union, or South Africa, where the arid zone constitutes a much smaller part of the general area.

Despite the fact that three-quarters of its area is arid or semi-arid, the standard of living in Australia is one of the highest in the world. This cannot be explained by cheap and inferior labour, for such does not exist here; the mineral resources are very important, but most of them have been discovered only in the twentieth century. The reason for the development is the population with its high technological level, which has been faced with challenges that required sophisticated technological and economic solutions.

The beginning of land use

Before the arrival of the first European settlers in 1788 there lived in Australia apparently some 300 000 Negritic Aborigines. The majority of this population of hunters probably lived in the humid zone, along the eastern coasts, where there were more hunting grounds than in the arid zone. The semi-arid region only began to be economically developed after the arrival of the European settlers.

The purpose of this migration was to create settlements of deportees, who would live on agriculture; commerce was not in view, and industry did not exist at all at that time. The forested ridge separating the coast from the interior had not been crossed before 1813, and the large-scale penetration into the drier interior started only after 1820. The environmental conditions were favourable to the development of sheep-breeding for wool, for which a ready, albeit distant, market existed in the expanding English woollen industry (Fig. 13.1). In those years, there was a steady outflow from agricultural areas in England, which were suffering from the effects of the enclosures, and by 1820 only half the Australian population consisted of the first deportees. The main populated area was in the south-

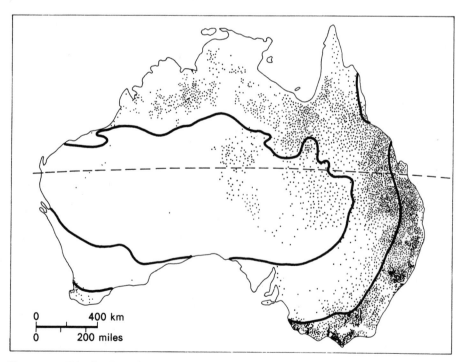

FIG. 13.1.   Distribution of sheep in Australia. Each dot represents 25 000 sheep (Wadham, 1961). The inner black line is the limit of the arid zone, the outer black line, that of the semi-arid zone.

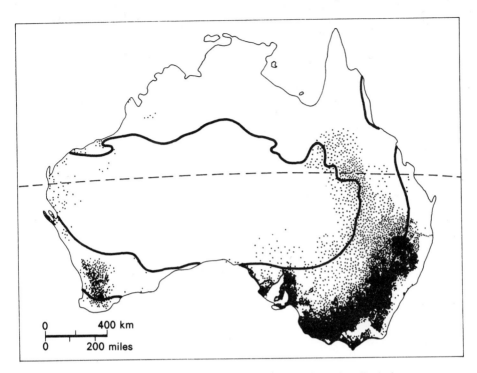

Fig. 13.2. Distribution of beef cattle in Australia (Wadham, 1961). Each dot represents 2 500 cattle. The black lines are the limits of the arid and semi-arid zones (cf. Fig. 13.1).

east: Sydney, Melbourne, Adelaide. The west's colonisation began in 1827. The east coast was comparatively poor in good soils; the good soils began behind the ridge. Most of the problems in sheep-breeding derived from shortage of phosphorus in the soil (as in South Africa), a detail the significance of which was appreciated only towards the end of the nineteenth century. Water was a limiting factor from the very beginning of colonisation. Settlement advanced mainly along the river valleys. Some lakes were too saline to be used for drinking water for the flocks; on the other hand water was found by simple drilling at the foot of the mountains, or in the beds of ephemeral streams. The developed area was expanding and spread only over areas with a considerably dense vegetation, except where there was no surface water; cattle were introduced into areas with denser vegetation, because of their ability to walk through it, and also at places far from water sources, since the animals can walk 15 km (9 miles) a day to a water source, as compared to 5 km (3 miles) that is the limit for sheep and goats (Fig. 13.2).

This process was slow and gradual. Further development was initiated, particularly by the artesian wells.

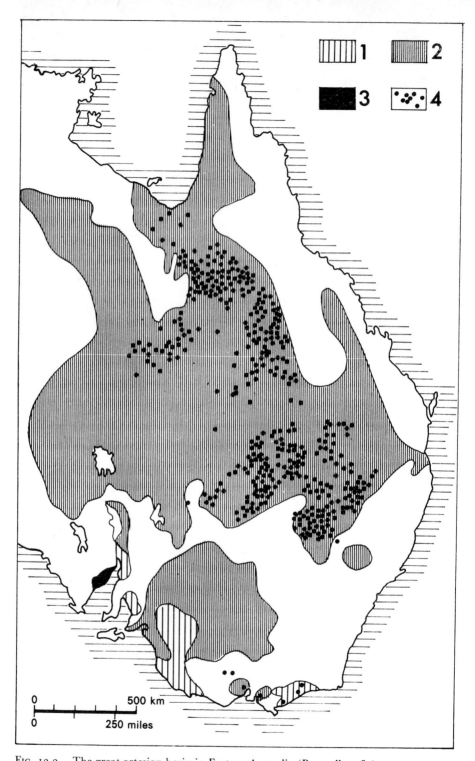

FIG. 13.3.   The great artesian basin in Eastern Australia (Pownall, 1967).
1. Water suitable for domestic use. 2. Water with light salinity suitable for livestock.
3. Water with unknown salinity. 4. Artesian borings.

The great artesian basin

The development of artesian wells as a water source started mainly in the Queensland arid area. In the 1960s there were about 8 000 wells with an average depth of 150 m (500 ft). There are also sub-artesian wells with a lower pressure. Figure 13.3 illustrates the large dimensions of the basin whose area covers a third of the whole of Australia. The artesian water is suitable for cattle, sheep and goats, but is too saline for crops; there is not sufficient fresh water in Australia, particularly in the arid regions: The artesian wells characterise the northern and the southeastern parts of the main artesian basin.

The first drilling in the great artesian basin was done in 1878 (Pownall, 1967), in the Kalara area; this drilling reached a depth of 50 m (160 ft); the column of water that spurted from the drilling reached a height of 9 m (30 ft). Two years later, in 1880, a series of droughts affected Australia, resulting in an accelerated search for water in the artesian basin, the nature of which only then began to be appreciated. Technological equipment was still poor; nevertheless, all the states in Australia, with Queensland leading, started underground drillings; Queensland alone produced in 1893 over 3·8 million cu m (5 m. cu yds) of artesian water a day. With technological progress, the drillings became more widespread to reach the distribution shown on Fig. 13·3.

The Beth-She'an area: a regional organisation for water supply on the border of aridity

The case of the Beth-She'an area, a triangle-shaped valley on the borders of the Jordan trough (Fig. 13.4), can be understood only against the background of Israel's agriculture, which is, nothwithstanding its limited proportions, one of the most dynamic agrotechnical enterprises in the world (Amiran, 1964). It also illustrates the complex nature of a settlement on the desert margin, by a population with an aspiration to a high standard of living, and the great variety of natural, technological, social and even juristic factors that contribute to the construction of such a society. As pointed out by Amiran, Israel's agriculture in general is characterised by three factors: it is based mostly on mixed farming, not on specialised branches, so that fruit plantations exist together with grain crops, etc; it is based, to a great extent, on cooperation of individuals in a certain village ('Moshav Ovdim', 'kibbutz'), as well as of the various villages ('Regional Councils'), and national organisations for marketing and purchase; it is assisted by enormous investments in mechanisation and irrigation. The intensification of agriculture by irrigation, use of fertilisers, introduction of suitable species and agrotechnical expertise, characterise the attitude of Israel. Out of a total 450 000 ha (over a million acres) of

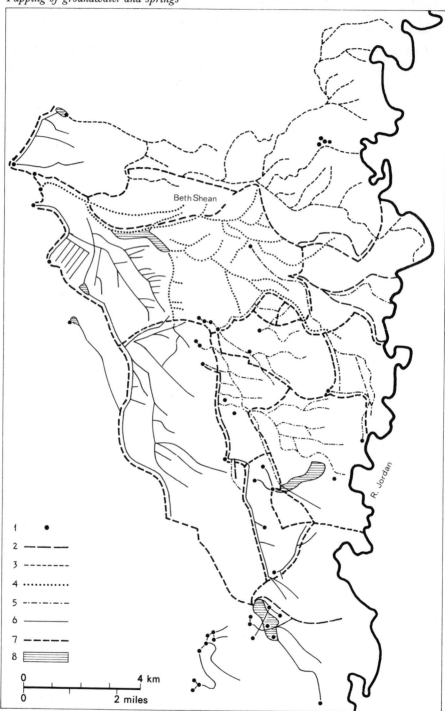

Fig. 13.4. Irrigation ditches in the Beth-She'an valley in the mid-1940s (Nir, 1968). 1. Spring. 2. Drainage ditch. 3. Ein Amal irrigation system. 4. Ein Migdal irrigation system. 5. Central level system. 6. Other systems. 7. Limit of irrigation basin. 8. Marshes.

cultivated land in Israel, about a third are under full irrigation.[1] All these factors are reflected in the individual case of the Beth-She'an area.

Among other factors affecting the agriculture in Israel, is the tradition, carried on from the first days of the renewed Jewish settlement in the country (by the movement of 'Returners to Zion' in 1882), which prefers agriculture to any other occupation. All the governments Israel has had until now have given priority to the needs of agriculture, in water, funds and development; the various agricultural organisations are affiliated to the Labour Party, which has been in power since the establishment of the state; this political factor has no small part in the success of agriculture in the marginal desert conditions which prevail in Israel in general, and in the Beth-She'an area in particular.

Physical conditions in the Beth-She'an valley

The administrative area under the jurisdiction of the regional council of the Beth-She'an valley includes an area of over 25 000 ha (62 000 acres) (Nir, 1968), most of which is a plain; the remainder comprises the Gilboa slopes and the plateaus of the Lower Eastern Galilee. Part of the soils are derived from Pleistocene lacustrine deposits, others were formed from tufa laid down by springs; generally, the soil is calcareous with a low percentage of organic matter, and its fertility is very limited without the addition of fertilisers. As for climate, this is a transitional area between the semi-arid zone and the arid zone; its southern border is the 180 mm (7 in) isohyet, and the rainfall increases to 320 mm (22·5 in) in the north. It is, then, a region on the fringe of aridity, where a large agricultural settlement could hardly be expected. Yet there exist here eighteen agricultural settlements (in addition to Beth-She'an city with a population of 12 000) which produce excellent crops. The reason for their existence is the great number of springs that are found along the western margins of the valley; these yield some 140 million cu m (183 million cu yd) a year, the largest discharge in the entire area of Israel, excluding the Jordan springs. About 40 per cent of these springs are saline, however, to the extent that they damage the cultivated plants (and the soil). The treatment of this brackish water, of the unfertile chalky soil, and of the pests which develop in such a hot climate (the area holds the record for maximum temperature in Israel, 56°C:132·8°F in June 1944), together with the problem of overcoming the distance from population centres, are features which make this region particularly interesting.

The tapping of water sources and their utilisation

The greatest number of springs is found in the northwestern corner of the

---

[1] This refers to the area inside the cease-fire boundaries of 1949.

Plate 16.   The water-mixing plant at Beth-She'an. The four canals are two pairs of two qualities (mixed and saline) of water (courtesy: the Director, Department of Geography, Hebrew University of Jerusalem).

valley where the elevation is highest; water could thus be delivered by gravitation to the other parts of the valley, even in remote times. On the other hand, the conduits were abandoned in times of wars or unstable government, resulting in swamps, malaria and the desertion of the area by the sedentary population. These situations alternated several times in the course of the region's history. The key, then, to the attractiveness or otherwise of the region lies not in nature, but in the hands of man.

The springs are a comparatively reliable source of water; in the main springs deviations from the annual average discharge are not more than 15 per cent. The small springs, on the other hand, which simply represent the reappearance of water that had percolated into the ground at higher levels, are unreliable and some of them have even dried up completely during the last decade or so. The first stage, then, in the utilisation of spring water, had to be research on their discharge and quality. In some of the big streams a mineral content (calcium, chlorine) of 1 000–2 000 mg/litre (570–1 140 mg/pint) has been found; before the present century this was either unknown to, or ignored by the cultivators, and probably many of the presentday saline soils are the result of long periods of irrigation with this saline water. The unequal discharges, the various qualities of water and the different rights to its use—according to the rules prevailing until the nationalisation of water resources in the first years of the existence of the state of Israel—made land use difficult until recently. When water resources were nationalised in the late 1950s a regional organisation was established in Beth-She'an, with the purpose of 'supplying every agricultural unit with an equal amount of water of the same quality'. To make this possible the water of different qualities from most springs is delivered to one place where it is diluted and then delivered onwards to the various agricultural settlements in an orderly and regulated manner (Fig. 13.5). This is one of the few instances in the world of irrigation where water with a mineral (chlorine) content of 500–600 mg/litre (280–340 mg/pint), is diluted with water containing 80–90 mg/litre (45–50 mg/pint), producing water with 250–300 mg/litre (142–170 mg/pint), chlorine content, which is appropriate for most crops; a dependable discharge is guaranteed, quarrels over water rights are avoided, and profitable crops can be planned. The organising of water distribution is not only an important technical enterprise, it is prim-

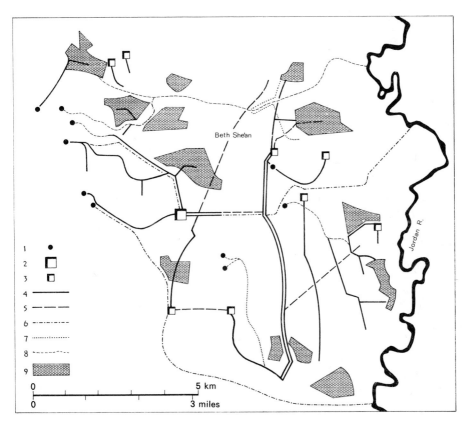

FIG. 13.5.   Irrigation ditches and water supply system in the Beth-She'an valley in the 1960s. (Nir, 1968). 1. Principal springs. 2. Regional centre of irrigation. 3. Water reservoir. 4. Irrigation canal (concrete). 5. Pipelines of irrigation. 6. Drainage ditches. 7. Irrigation canal (earth). 8. Wadi. 9. Fishponds.

arily a juristic and socio-economic undertaking, underlying all other economic activities in the region.

Today the cultivated areas receive between 8 000 and 20 000 cu m (11 500 and 706 000 cu ft) per ha according to the type of crop; the water allocations are larger here than in other parts of the country, since evaporation is more intense, and rain has small practical value as an irrigating factor. Even wheat and barley must be irrigated several times during their period of growth. Because of the high value of water, special consideration is given to irrigation methods; the conduit ditches are all cemented. Inundation is not practised; instead, irrigation is performed from levelled ditches or by sprinkling; the latter, however, is carried out only at night, because of evaporation and winds that prevail during the day, annulling the effect of irrigation by this method.

Yet the water problem is not the only one dictated by nature in this marginal region. Soils are either chalky or sterile, or hydromorphic (swamps in areas that were inundated by spring water through neglect of former generations). The treatment of chalky soils is expensive but practicable, and today these soils produce large yields of alfalfa, cotton and sugar beet, that are processed in local factories. The swamps have been turned into fishponds, and these areas which seemed lost for agriculture, are the most profitable ones, since a hectare of fishpond gives a higher return than any other kind of agricultural land. Here, too, the problem of having to use expensive water does not exist, since the water used for the fishponds is

Plate 17.   Fishponds in the semi-arid landscape of Beth-She'an valley. The water used here comes from saline sources (courtesy: the Director, Department of Geography, Hebrew University of Jerusalem).

brackish (approximately 1 000 mg of chlorine per litre (570 mg of chlorine per pint)) and could not have been put to any other use.

From the foregoing it is apparent that one of the gravest problems in the beginning of settlement was the need for a deep and ramifying agricultural drainage. At the present time the groundwater sources are well under control; the surface runoff, coming in floods from the slopes of the Gilboa mountain, had also to be confined in ditches to prevent inundation of fields and settlements. Another inherent climatic problem is the rapid development of pests: the *Prudenia*, a caterpillar which attacks the leaves of the cotton plants, develops in this hot climate three times more rapidly than in the Mediterranean regions of Israel; pest control measures have therefore to be taken more frequently. Malaria, which during the 1940s claimed many victims, no longer exists; it has been abolished not merely by draining the swamps, but also, and primarily, by insecticide spraying, a fact which is expressed in the regional council's budget. Under these natural conditions—or challenges—there exists today an agricultural population in eighteen villages who make their living from agriculture. The regional organisation is highly developed: the settlements have been incorporated in an economic society, which acts through institutions that would have been beyond the powers of a single settlement; the distance from other centres of population has also been a factor in this regional organisation. In addition to its responsibility in the matter of water distribution and the planning of the various branches of agriculture, this organisation also controls factories for the initial processing of agricultural products. One of the first enterprises was the factory for drying alfalfa; the first gin in Israel for separating the cotton fibres from the seeds was built here (the Beth-She'an valley was also the first to grow cotton, with great success); in the course of time, a regional slaughter house for fowls was built, providing the market with plucked, packed and frozen fowls; there is also a factory for canned dates, olives and various vegetables; in fact, the regional centre regulates all the agricultural activity of the region, holds heavy equipment for earth works, and organizes the pest-control operations. It is hard to imagine the economic activity of the region today without these cooperative enterprises.

Today, nearly 15 000 ha (37 000 acres) are being cultivated in the Beth-She'an region (Nir, 1968); the major part of this, over 12 500 ha (32 000 acres) are in kibbutzim, and the remainder in moshavim.[1] Cultivated

---

[1] A 'moshav' is a rural settlement on national land, where each family cultivates its plot accorded for a lifetime. The production and consumption are entirely the private business of the family, but there are certain rules as for the limiting of hired hands; there exists also an extended cooperation in the marketing of the products and in purchasing of materials.

A 'kibbutz', being also a settlement on national land, is entirely based on collective work and consumption; there are no private plots, and the economy is entirely centralised by the kibbutz. The private property of its members is limited to their personal belongings, and there is no private ownership of the means of production. The kibbutz is also highly industrialised, sometimes up to 60–70 per cent of its total production.

pastures cover 1 100 ha (2 700 acres), dry farming nearly 5 000 ha (12 300 acres) (37 per cent) and the rest is in irrigated fields. Of the 'unirrigated' crops, only a third are actually not irrigated at all, and the rest are sustained by several waterings during the growing season. The major part of the irrigated land (51 per cent of the entire cultivated area) is devoted to industrial crops (cotton, sugar beet, alfalfa); second come fishponds, which occupy 11 per cent of the total cultivated area; fruit plantations constitute only 8 per cent, a much lower proportion than in other parts of the country. Vegetables are not given priority and constitute only 4 per cent of the entire cultivated area.

Kefar Rupin, a kibbutz on the Jordan banks which suffered greatly from shelling by terrorists during the years 1968–70, may serve to illustrate the diversified character of an agricultural settlement in the Beth-She'an region. Its cultivated lands are divided as follows:

Land use	percentage
Cereals	35
Fodder	29
Fruit trees	6·5
(particularly dates and pomegranates)	
Cotton	6·5
Sugar beet	5·5
Cultivated pasture	5
Fishery	12·5
	100

The diversity of agricultural production permits an even distribution of labour through the year, though there are, of course, seasons of more intensive, as well as periods of less intensive work. Fig. 13.6 shows that in the settlement's agricultural calendar the principal season is autumn, when the harvesting of olives, dates and cotton is done simultaneously, whereas the winter months are mainly the growth season.

FIG. 13.6   Agriculture work schedule in Kefar Ruppin in the Beth She'an valley, 1958 (Nir, 1968). 1. Grain. 2. Fodder. 3. Orchards. 4. Fishponds. 5. Dairy cattle. 6. Beef cattle. 7. Cultivated pasture. 8. Poultry. 9. Nutrias.[1] 10. Sugar beet. 11. Cotton. 12. On the ordinate, number of workdays in one month; on the abscissa, the months.

[1] Nutria (*Myocastor coypus*), a small rodent of South American origin, was introduced in the 1950s into the fishponds for dual purposes: as a rodent, the nutria was most effective in clearing the banks of the ponds from the invading reeds; as a fur-animal, its furs were even exported abroad. The raising of nutrias declined in the middle 1960s.

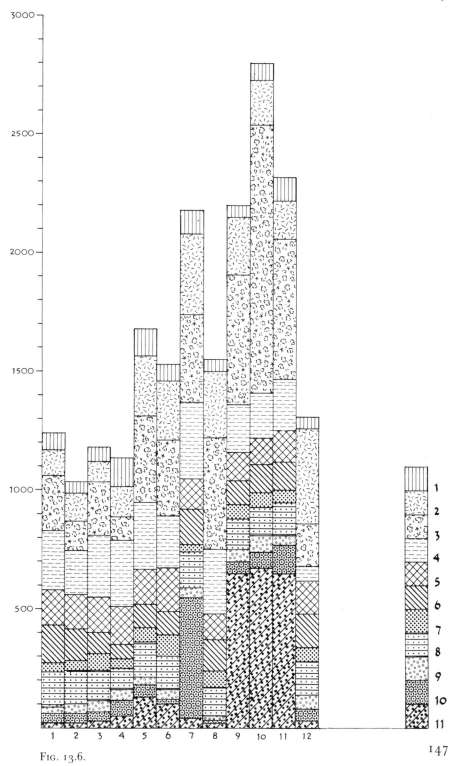

Fig. 13.6.

The agricultural economy has adjusted itself to the special conditions of the region: a long summer, a mild winter and a short spring. Different agricultural products are given prominence here, compared with the Mediterranean parts of the country. Wheat and barley are mostly grown with supplementary irrigation; the region adopted this policy after a series of droughts during the years 1958–62, when most of the wheat and barley fields were severely affected (Nir, 1963). In 1958–59 rain amounts in the Beth-She'an valley ranged from 65 mm (2·5 in) in the south, to 90 mm (3·5 in) in the north, and of 4 300 ha (1 000 acres) which had been sown, 3 800 ha (9 400 acres) yielded no harvest. In other parts of Israel (excluding the northern Negev) crops did not suffer at all from droughts during those same years. Those years of drought induced the formation of a special fund which is operated jointly by the government and the farmers, for covering the losses from drought. Vegetable-growing is also different here from other parts of the country: while the main seasons for vegetables, over most of Israel, are autumn and spring, in Beth-She'an, because of the short spring, the growing season is the autumn, and no vegetables at all are grown during the hot summer.

Seasons of vegetable growing in Israel and in Beth-She'an Valley, 1959 60
(percentage of total)

	Autumn	Winter	Spring	Summer
Israel	32	21	33	14
Beth-She'an	65	19	15·8	0·2.

The question arises, what is the contribution of the Beth-She'an region to the agricultural economy of Israel? The cultivated area constitutes 3·3 per cent of the entire cultivated area of the country; but the specific contribution of various crops accentuates the exceptional nature of the region and the importance of its contribution to the national economy. The wheat grown in the region constitutes 6 per cent of the total wheat production of the country (1960); this is twice its proportional share according to area, and shows the priority given to this crop. The production of fruits, vegetables and dairy cattle does not reach 2 per cent of the total national production in these branches, but on the other hand, the region's share of cotton and sugar beet is 10 per cent of the national total, a contribution surpassing expected proportions; the area of fishponds—18 per cent of the total inland fisheries in the country—gives the Beth-She'an region prominence as an important producer of fish, which have always been an important source of proteins for the population.

This region thus has a very special agrohuman character: in the middle of hostile natural conditions a population with a well developed agrotechnical expertise has arisen, attaining considerable agro-economic achievements and an appropriate standard of living. The advanced regional organisation, the adaptation of agriculture to the special conditions, which are not always adverse (for instance, growing early vegetables for export to Europe, or early grapes that reach the west European table in May), the concept of economic life as a constant endeavour and a continual improvement of methods of cultivation and maximum employment of the latest agrotechnical achievements, all these may be considered as contributing factors. Man's domain of occupance has thus been expanded and the desert has retreated southward.

14
Surface water

Most important hydrotechnical projects have been established on great rivers—the Nile, the Hwang Ho, the Indus, the Colorado, to mention only the most outstanding; they generate extremely intensive patterns of life which owe their whole existence to water. Yet the majority of these projects are outside the semi-arid zone. The following pages will examine a few regional projects which have radically changed life patterns in the semi-arid zone without upsetting other forms inherent in the semi-arid environment.

River water in northwestern Mexico

Mexico is almost entirely an arid country; only 17 per cent of the country is humid (Streta and Mosino, 1963); 36 per cent is arid, and 47 per cent semi-arid. Northwestern Mexico, during the first half of the present century, was the arena of a combined activity involving the conditions of the semi-arid environment, irrigated agriculture and agrarian reform. This was an advantageous combination; yet several dangers appeared in connection with land use and methods of cultivation. The beneficial results were obvious: the development of the irrigated lands included in the agrarian reform turned them into the most productive area in the country (Henderson, 1965).

The agricultural lands of northwestern Mexico are located by the deltas of the rivers (Fig. 14.1). In this area the rain is not reliable enough for agriculture: the amounts of precipitation range from 75 mm (3 in) in the north, by the Colorado delta, to 400 mm (15·7 in) in the south, by the Fuerte delta. Irrigated lands are found in this area by the Colorado delta, in the Caborca region, on the Magdalena river, in the valley of the Sonora river, and particularly on the deltas of the rivers Yaqui, Mayo and Fuerte. These three rivers have added to the narrow coast plain of northwestern Mexico large deltaic areas, with fertile alluvial soil; these plains are exceptional in a landscape which is mostly steep and hilly. The government of Mexico has constructed since 1940 three dams, one on each river, regulating the flow by preventing floods and supplying water for the irrigation of the delta lands. This project is one of the greatest, not only in Mexico but in the whole world

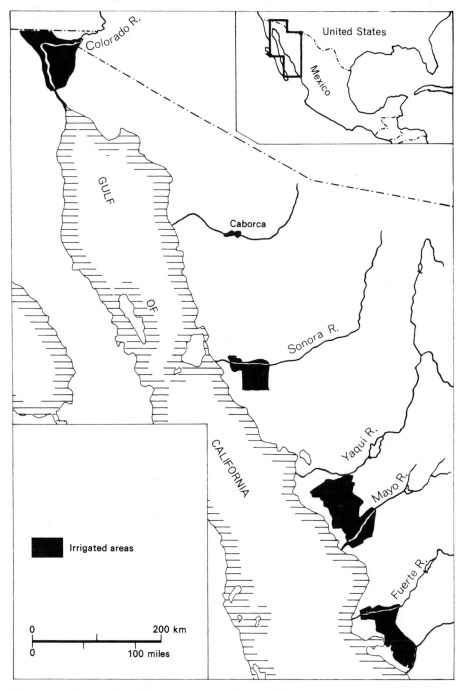

FIG. 14.1.   Irrigated lands in northwestern Mexico (Henderson, 1965).

(Dozier, 1963), and the area it covers is vastly greater than the original riparian lands which existed as oases even in pre-Columbian times.

The cultivated area, that was divided among the farmers as part of the agrarian reform, was mostly virgin land; many development works were necessary, such as digging irrigation and drainage ditches, building embankments, and constructing roads; all these, including water supply, have been organised by the governmental agency which is in charge of the reform.

Much of the land is now owned collectively in 'colonies'. A private owner is permitted to hold no more than 100 ha (250 acres); but cotton growers are permitted to own up to 150 ha (375 acres) and sugar beet growers as much as 300 ha (750 acres).

The history of intensive cultivation in northwestern Mexico is short, beginning, in fact, during the 1930s, when the agrarian reform was launched. In 1926, the irrigated area covered no more than 150 000 ha (375 000 acres); by 1963 it had grown to 850 000 ha (2 100 000 acres). A small portion of the land was already cultivated by the native Amerindian population. Also, on the Colorado delta there were irrigation works dating from 1904, owned by American companies; these lands were taken over during the agrarian reform in the 1930s.

The rapid development and redistribution of lands in the years 1934–40, beneficial though it was, nevertheless raised several problems, the most important being those associated with cultivation methods and the conservation of soil fertility; these technical matters had been neglected in the process of land apportionment, which seemed at the time to be the principal issue. Water was not adequate, in most cases, for the needs of an economy that was developing so rapidly. Thus, for instance, the irrigated lands in the Colorado delta had to be reduced to 190 000 ha (470 000 acres), since the total amount of water was not sufficient for irrigating the entire area that had been divided among the poor during the reform.

In the Questa de Hermosilio, on the Sonora river, 350 million cu m a year, provided by bore holes, are used to irrigate 130 000 ha (321 000 acres) but the amount extracted has been three times as much as the annual natural replenishment of the storage, and the continuity of the supply has thus been endangered. Here too, it has been necessary to reduce water consumption by 40 per cent.

The dams on the Yaqui river are capable of holding water for the irrigation of 250 000 ha (618 000 acres) but in this case it is necessary to limit the irrigated land because of the alkalinity of the soil. But in the majority of cases the water stored behind dams permits a more effective use of the cultivable land.

The problem of preserving soil fertility exists in any arid region where the drainage is not well organised, and it becomes even more acute in places where irrigation is the dominant method in cultivation. As in most arid soils, here in Mexico the organic content of the soil is no more than 1 per cent.

Generally, fertility has decreased since the commencement of cultivation, mostly through salination. In the valley of the Yaqui river, 40 000 out of 240 000 ha (98 800 of 593 000 acres) have already been affected by salination, and 15 000 ha (37 500 acres) have been taken out of cultivation. In the Colorado delta 14 per cent of the entire agricultural area has been affected by salts.

There is a distinct connection between soil fertility and the form of land ownership brought about by the agrarian reform. The main concern of the authorities, at the time of launching the reform, was with the distribution of lands and the organisation of water supply; there was no time, or the necessary money, for making equal efforts in soil drainage. The groundwater table in this deltaic region is high; the farmers themselves did not arrange such a drainage system, either from lack of expertise and money, or from uncertainty as to the future ownership of the land. There is a constant danger of rapid soil salination because of the flat and level land surface and nearness to the sea. Only a thorough draining and pumping out of excessive saline water from the soil may improve the situation. Another reason for decreasing soil fertility was the absence of fertilising and deep ploughing. All these matters improved when the reform and the changes in land ownership were more firmly established.

The financial problems in the new lands are no less acute than the technical problems of irrigation; since cultivation is mechanical, the farmer has to buy, or hire, agricultural machines. The settling agencies advanced loans on the strength of forthcoming harvests. Very few long-term loans were granted, and the interest on them was rather high, at 10–18 per cent. The high interest forced some of the settlers to use for subsistence money that had been intended for the purchase of fertilisers or pest control materials. Thus a vicious circle was formed: when the lands fail to yield the expected harvest the debts grow, there is no way to get more loans and the debts continue to accumulate. Most of the loans are given for certain crops only: wheat, cotton, sugar beet; part of the harvest is bought by the government for fixed prices, which are guaranteed in advance.

The development of the northwestern region in Mexico was made possible by the construction of transport facilities which enable the crops to be marketed in other parts of Mexico and in the United States, and by the building of ports for cotton export to Japan and to Europe.

After more than a dozen years the region proved its advantages: the Yaqui valley was in 1960 the most important wheat-growing district in Mexico, and in the same year over 45 per cent of Mexico's total cotton production was grown in those northwestern regions. The Fuerte valley became the most important sugar beet growing district in the country. Rice is now being grown on the new lands, where the yields have been greatly increased. Overall, these northwestern districts constituted 30 per cent of the total irrigated land in Mexico. The crop yields were higher than the average of the other irrigated lands (Dozier, 1963).

Summary

The development of the semi-arid lands in northwestern Mexico has proved, in two decades, that a combination of several factors—climate, irrigation, organization of land ownership, governmental control and guidance—can raise to the position of one of the state's most flourishing agricultural centres, an area which is marginal with regard to its population density and its productivity. In the deltas of the rivers Yaqui, Mayo and Fuerte there were in the 1960s 34 000 farmsteads, on an area of 100 000 ha (247 000 acres). In addition to the rural population, there was also, in 1960, an urban population of 120 000. Here too, the specific problems of the region were encountered; in this case it was found that the drainage problem had previously been neglected, and was consequently amended. The financial system also involved some problems.

The size of the farms allocated to private, or collective owners varied from medium plots of 20 ha (50 acres) to large plots of 300 ha (740 acres). This pattern was found to suit the needs, and plots of these measurements could be machine-cultivated, thereby reducing the costs of production and making competition in distant markets feasible. Again, without a system of roads, railroads and ports, and without marketing organisation, the policy of planning agriculture on the lines of a market economy could not have succeeded.

It is apparent that this fundamental change in the semi-arid landscape of northwestern Mexico is not the result of a single factor (irrigation) or of the agrarian reform. It is a good illustration of how, by regional planning and the cooperation of national organisations which acted convergently toward the common goal, the desired end has been attained.

Changes of values in the Saragossa basin affected by the Ebro river's water

The semi-arid zone in Spain

According to precise definitions of arid zones, 1·5 per cent of Spain may be defined as arid, since it receives less than 300 mm (11·8 in) rain a year; but 36·5 per cent, more than one third of the country, receives between 300 and 500 mm (11·8 and 19·7 in) rain, and may thus be defined as semi-arid. The limitations on crop types will be severe in this large area, and without artificial irrigation, it will be hard to guarantee a proper standard of yields (de Labura, 1964).

Only a third of the total irrigated area in Spain (1·8 million ha [4·5 million acres] in 1960) receives its water from underground sources; the remainder comes from storing surface water, particularly river water coming down from the Pyrenees, the Cantabrian massif or the western plateaus. This water is of excellent quality. The irrigation enterprises are built by the government, which is also concerned with land settlement schemes, agrarian reform and land consolidation.

Enterprises in the basin of the Ebro river before the present century

The irrigated lands of Spain have undergone since the beginning of the present century, a radical change. While in the early years of the century the principal irrigated lands were in Andalusia and the Levante coast, in the 1960s the majority of the irrigated lands were in the valleys of the great rivers, particularly the Ebro, which now contains nearly 30 per cent of the total irrigated lands of Spain. This is an expression of the action taken by the Spanish government for the settlement of landless farmers, a large-scale development enterprise (Bomer, 1964).

The valley of the Ebro river (Fig. 14.2) is suitable for big irrigation projects, by reason of its physical character. It is surrounded by high mountains, especially in the north and northwest (the Pyrenees and the Vasco–Cantabrian massif), that receive great amounts of rain and snow; but the core of the basin between Lérida and Saragossa gets less than 300 mm (11·8 in). Rain in the Ebro basin, as in any semi-arid region, varies greatly from one year to another. In this region the existence of the olive trees and even the vines is marginal, and dry farming, too, is a 'hazard agriculture', without supplementary irrigation. Only one year in seven is a 'good' year, two are middling and four are liable to drought. Such conditions confine settlement to the banks of the streams, and the number of settlements in the *secano*, that is, the unirrigated area, is very small, and even these have been neglected because of the undependable economic basis.

The structural axis of the Ebro valley and the course of the river, which are more or less perpendicular to the streams coming down from the Pyrenees, turn the Ebro into a large conduit which gathers the water from all the rivers into one channel. The greatest amount of water arrives from the north—the Pyrenees; when the river enters its central basin, its discharge already amounts to 80 cu m (105 cu yd)/sec; the Aragón joins with 160 cu m (210 cu yd)/sec; then the Gállego with 40 cu m (52 cu yd)/sec; and finally, the united Segre and Cinca, flowing into the Ebro near Fraga, with a discharge of 200 cu m (260 cu yd)/sec. But the discharge of these rivers decreases in summer, and it is necessary to build dams for storing their water before they enter the Ebro. Also, these river valleys are deeply cut down in the piedmont which is built of non-resistant Tertiary deposits, so that they sometimes flow at a depth of several hundred metres below the piedmont level. A large part of the Tertiary rocks which fill the basin consists of marl. Water delivery in previous centuries was rendered difficult by these conditions. Water enterprises were first built here towards the end of the Middle Ages, and they continued to operate in the sixteenth and eighteenth centuries; but the discharge of those canals—for instance, that of the 'Imperial Canal', which was built in the sixteenth century, ranged between 25 cu m (33 cu yd)/sec during the wet season and 5 cu m (6·5 cu yd) /sec in the dry season. Many of these enterprises were destroyed by the collapse of the marl.

Development during the present century

The organisation of water delivery and irrigation received its great impetus in the beginning of the present century (Fig. 14.2). Above Tudela, the Ludosa canal was built on a terrace of the Ebro river; its construction began in 1915 and was completed in 1936. It is 125 km (78 miles) long, its discharge can reach 22 cu m (29 cu yd)/sec and it is capable of irrigating 25 000 ha (61 775 acres). The dam on the upper Ebro near Reinosa holds over 500 million cu m (650 million cu yd) of water. The majority of rivers flowing down from the Pyrenees—Aragón, Cinca, Gállego, Segre—are equipped with dams built at points where the river emerges from water gaps in the mountains that fringe the Pyrenees; the Aragón irrigates 62 000 ha (155 000 acres); the Gállego, about 150 000 ha (370 000 acres) in the south; the Cinca will irrigate, after the completion of a canal now under

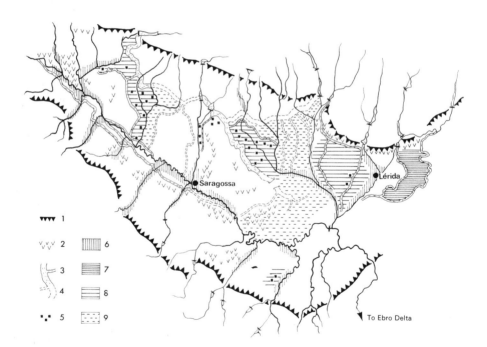

FIG. 14.2. Irrigation in the Ebro river valley, Spain (Bomer, 1964).
1. Mountain border. 2. Gypseous soils. 3. Canal in function. 4. Canal in construction. 5. New settlements. 6–9 irrigated lands: 6. Ancient; 7. Irrigation in eighteenth and nineteenth centuries; 8. Irrigated during twentieth century; 9. Irrigation projected.

Plate 18. Puellatos, a new village on the Gallego Canal, Ebro basin, Spain (courtesy: the Director, Department of Geography, Hebrew University of Jerusalem).

construction, an additional 50 000 ha (123 000 acres). The total sum of the additional land which is irrigated by the new enterprises is over 325 000 ha (803 000 acres). The cultural landscape of this enormous area has changed completely.

The new settlements

The traditional agricultural landscape of the Ebro basin has been changed only recently, by the enterprises of the present century. Though considerable parts of the river terraces have been irrigated for over a hundred years, the traditional crops—wheat, vines and olives—dominated the agricultural economy. The absence of extensive progressive changes may be attributed partly to the other climatic handicaps, such as the frost that frequents the region. But the social structure and particularly the distribution of land property, also held back progress. Only a change in the size of land property could give impetus to colonisation. Thus, for instance, in the Saragossa district, the National Institute for Colonisation[1] (which is a government agency in charge of the colonisation activities in the region) bought, among other things, two farmsteads from their former owners; the

[1] Now known as Instituto Nacional de Reforma y Desarrollo Agraria.

157

farmsteads were large enough for the establishment of two villages, Subradial and la Juyuza, today populated by several hundreds of families.

The colonisation on the irrigated lands, the *huertas*, causes a decrease in the *secano* population. Moreover, *secano* lands are turned into *huertas* by aid of the irrigation ditches. In order to realise the colonisation enterprises, the National Institute for Colonisation began to reduce the size of ownership units by the purchasing of lands and their subdivision among the settlers of the new villages. Before this reform started, 75 per cent of the *secano* lands in certain regions had consisted of plots bigger than 100 ha (247 acres); in 1964, a farmer could keep no more than 30 ha (74 acres) of land—which had been converted in the meantime into irrigated land. The Institute for Colonisation prepared the land for irrigation, by constructing a road system, levelling the ground and building the basic stuctures in the new villages. These villages have from 500 to 1 000 inhabitants. The houses are built in accordance with the regional tradition, with an enclosed courtyard surrounded by a wall; only the front of the house touches on the street. Twenty such villages have been built in the Saragossa district. The area allocated to a new farmer's family in a village of this type is 7 to 10 ha (17 to 25 acres), all under full irrigation; there is an unlimited supply of water, but the farmer has to pay for the amount used. The Institute provides the farmer with tools on credit for five years. If the farmer wishes he may purchase his land, paying in small instalments for thirty years; as long as he does not do this, he is dependent on the Institute for Colonisation, and is, in fact, the Institute's tenant. The lands may be passed on in inheritance, but not sold or divided into subplots, so as to prevent one of the traditional village handicaps, namely, the reducing of the area in the possession of a family. Along with these socio-economic changes there occurred essential changes in crop types: the general tendency is to pass from subsistence agriculture to market agriculture, and the new crops are cotton, fruit trees, and breeding of dairy cattle. Marketing is mostly cooperative.

Summary

This experiment in intensification is new for Spain, and is faced not only by economic difficulties, but also by social and psychological difficulties. Yet the experience of other countries where new farmers have been put on new lands shows that the results are usually favourable, and similar results may be expected in the Ebro basin. The experiment also shows (at least in Bomer's view) that the nature of the government is not a determinant in the organisation of the area for increasing production, and that ideas on agrarian reform are not the exclusive property of Socialist governments.[1]

[1] Spanish geographers would not agree with this view. See also Naylon (1965, 1973). (Editor's note.)

The semi-arid zone in Pakistan

The semi-arid zone in Pakistan is part of the arid belt extending from the Atlantic Ocean, through Sahara and the Arabian Peninsula, Iran and Afghanistan, to the Aravalli range in central India. This is the least populated area in the two most densely populated countries; still, tens of millions of people live here.

This arid region includes a belt of a more humid climate, which is semi-arid because of winter rains (Fig. A3). The Indus tributaries and the Indus itself also enable irrigated agriculture to exist in this semi-arid region. Apparently the culture of irrigation was present here in prehistoric times; in modern times the British started to build diversion dams on the Indus, as early as the middle of the nineteenth century (Bharadwaj, 1961); the large enterprises were started early in the present century and during the 1930s. In the semi-arid area in Pakistan and northwestern India, within the Indus basin, there are now about 8 million ha (20 million acres) of irrigated land.

Basic conditions

The greater part of Pakistan, excluding the southern Himalayan slopes and the adjacent piedmont, is located in the arid and semi-arid zone. The semi-arid zone includes the central part of the Punjab and the western mountains (Ahmad, 1964). In this zone the soils are mostly alluvial, and the land is almost flat, with a moderate gradient. The surface can be divided according to its elevation above the river channels, into the active channel, the periodic flood plains, and above them two river terraces, which are the ancient river floodplains. In the west, at the foot of the Suleiman mountains, extends the sandy undulating surface of the piedmont. The most westerly part, in Baluchistan, has a hilly and dissected landscape; as for the availability of water, this is the region of the underground qanats, belonging to the inner Asian basins, like Afghanistan and Iran (page 64). Rainfall in the semi-arid zone ranges between 250 and 500 mm (9·8 and 19·7 in), and is concentrated in the summer, though there is a small amount of rain in winter, too. In the west most of the rain falls in winter. Northward, the late winter and early spring rains assume greater importance; the snow in the northwest supplies a constant source of groundwater, that feeds the qanats. The annual variability of rainfall is as much as 30 or 40 per cent (Fig. 1.2).

Settlements in the semi-arid zone

In Pakistan the rural population in 1961 constituted about 77·5 per cent; but the 22·5 per cent of urban population were largely the product of intensified urbanisation that occurred during the years 1951–60, concurrently with the obtaining of independence and absorption of Muslim immigrants from India. Urbanisation, which was also associated with increased industrialisation, attracted a flow of manpower from the rural areas into the cities. In the semi-arid region there are five big cities of over 100 000 inhabitants; Lahore itself is a city of a million and a quarter (1961) one of the biggest cities in the semi-arid zone. The rural population is dense in the east, and becomes more scattered towards the west; in the western mountains there are settlements only along stream courses.

Types of agriculture

Figure 14.3 illustrates the ratio of cultivated areas to uncultivated areas, and also the dry farming and irrigated lands. As might be expected, the western mountainous parts, between the Indus and the Afghanistan border, are mostly uncultivated; even the part that is cultivated is mostly irrigated, and there is almost no dry farming. The situation in the southeast of the semi-arid region is similar: its major part is irrigated. On the other hand, the northwest, which enjoys a certain amount of precipitation even in winter, includes large areas of dry farming that are often several times larger than the irrigated lands. The amounts of water for irrigation are

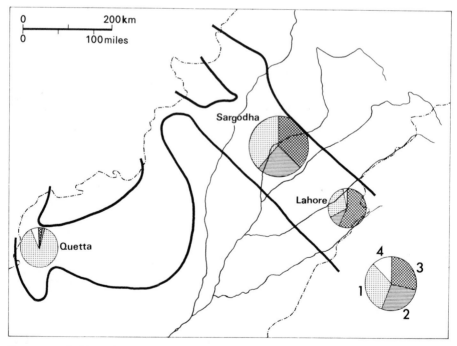

FIG. 14.3. Cultivated lands in selected districts in the semi-arid zone of Western Pakistan (Ahmad, 1964). 1. Uncultivated. 2. Cultivated (unirrigated). 3. Cultivated (irrigated). 4. Area not reported. The black lines delimit the semi-arid zone.

determined, of course, by the nature of the water sources and the methods of delivery. Generally, the amount of water depends on the seasonal behaviour of the rivers. Because of differences in discharges according to seasons, it was necessary to construct dams. There are two types of ditches: permanent and seasonal irrigation ditches.

Perennial irrigation exists in the Punjab, in the interfluves between the five tributaries of the Indus. Here, winter crops prevail, with wheat as the principal crop. Cotton is grown in summer; the all-year-round irrigation turns this part into the most fertile area in the whole country.

Seasonal irrigation exists on the ancient river floodplains; the principal season for irrigation is summer. In addition to the use of surface water, there are wells drawing from the groundwater table which is here close to the surface.

An illustration from the Quetta region

The Quetta region is located in the western part of the semi-arid belt of Pakistan (Figs. 14.3, 14.4); its area is about 380 sq km (147 sq m) (1965), most of it mountainous; only 2·5 per cent of the area in the alluvial flood-plains along the streams is cultivated. The climate is particularly severe: there are regular frosts in November-February, whereas the summer is very hot. Winter is the rainy season; there is a great variability of rain amounts from one year to another. Thus the annual average is 220 mm (8·66 in falling on 44 days), but in 1925 there were only 85 mm (3·35 in) of rain, and in 1885, 540 mm (21·2 in) (Dehman and Verstappen, 1965).

Agriculture is based on irrigation in a few qanats, but mainly on flood water; the first method is applied to the fruit groves and the vineyards, the second to wheat and barley fields. There are villages in which 65 per cent, and even 91 per cent of the cultivated land is under irrigation, mainly by the quanats; cornfields occupy up to 80 per cent of the total area. In some cases, up to 25 per cent of the fruit harvest is lost through the spring frosts; on the other hand, sown fields are sometimes destroyed by the winter floods.

An illustration from Ludaka and Senda in the southeastern part of the semi-arid zone

The village of Ludaka is located in the centre of the semi-arid region, on an interfluve between tributaries of the Chenab river. The proportion of cultivated land is very high (83 per cent) and all of it is irrigated by ditches. Over half the cultivation (55 per cent) is of market crops, due, to a great extent, to the proximity of urban centres; the principal crops are cotton and wheat. In detail, the proportion of the land under the various crops is as follows (per cent): wheat 24, fodder 23, cotton 14, millet 14, and about 5 per cent for each of the following: maize, fruits, vegetables, oil plants and sugar beet.

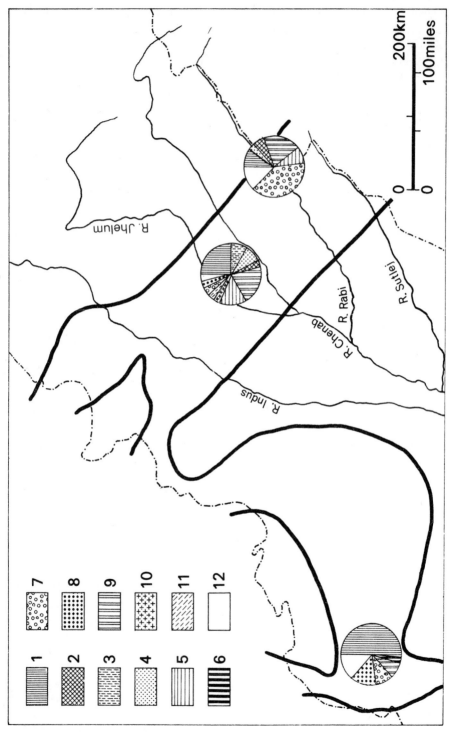

Fig. 14.4.   Types of crops in selected villages in Pakistan (Ahmad, 1964).
1. Wheat. 2. Maize. 3. Chickpeas (grams). 4. Millets. 5. Cotton. 6. Barley.
7. Vegetables. 8. Fruits. 9. Fodder. 10. Sugarcane. 11. Oilseeds. 12. Others.
The black lines delimit the semi-arid zone.

The village of Senda is located on the flood plain of the Ravi river, on the Indian border, within 3 km (less than 2 miles) of the city of Lahore. Rainfall is less than 500 mm (19·7 in) the greater part of the water is delivered by ditches, or drawn from the groundwater table which is close to the surface. The cultivated land occupies 65 per cent of the village area; 56 per cent of it yields two harvests a year, one in summer and one in winter. There is no unirrigated land here. The proximity of the big city has a striking effect: three-quarters of the cultivated area is occupied by market crops. This means that the farmer has cash, and therefore a certain purchasing power; he can make improvements and raise his standard of living. The fields are small, but self-sufficient, due to intensive cultivation and to the marketing of produce. The principal crops, according to the area they occupy, are as follows (percentages): vegetables for the nearby city markets, 43; fodder (for dairy cattle, also for the market) 24; wheat, 10; maize and cotton—8 each; and a little sugar cane.

Summary

The semi-arid region in Pakistan has great potentiality because of the large amount of available water, and the soils, which are mostly alluvial. The urban population constitutes a large potential market, which is likely to increase in importance in the future, as the rising standard of living enables the town-dwellers to purchase agricultural products in greater quantities. Even now there is a distinct difference between an area which is distant from market centres, like the Quetta district, where traditional life continues, and an area which is close to urban centres and can grow market crops, thereby developing a comparatively high level of technology and economy. The region is apparently on its way to commercialising its agriculture, which is the first step towards development.

15
Summary: the lesson from the detailed studies

The problems of arid countries have now been diagnosed, and the various reactions of mankind to the challenges have been outlined and assessed. It remains to summarise the conclusions reached.

The basic conditions common to all the semi-arid regions are known: lack of precipitation, lack of soils, abundant insolation; similarly, the conditions that are not shared equally by the countries within the zone are also known: the existence of underground water; allochthonic rivers; water storages in nearby snow-covered mountains. Basic anthropogeographic elements whose distribution over the semi-arid countries is even more irregular, are also known: ethnic compositions, political structures, distances from marketing centres, and the penetration of education.

The processes that can bring solutions to the problems of the semi-arid zone are also known: facilities for water damming, deep drillings, use of solar radiation for creating energy, creation of tourism and recreation attractions, air-conditioning, water desalination, extension of transportation, adaptation of crops.

Knowledge of the conditions and possibilities is thus available. The question is, what turn do affairs actually take? The detailed studies which have been presented in chapters 11–14 give some notion of the interplay between the various factors and prospects.

It appears that as long as a certain settlement persists in the traditional way of life and does not adjust itself to the modern economic environment —though this same way of life is sometimes extraordinarily well adapted to the natural physical environment—there is not much chance for better conditions of life for future generations. The isolated form of life, without contacts with a developing economic hinterland—preferably an industrial-urban hinterland that requires food supply, recreation space, vacant land for developing industry— has not proved itself capable of developing the semi-arid regions by its own powers, and raising them to twentieth-century standards of living. Samples from countries which differ greatly, make it clear that without action in regional, national or global frameworks, full and profitable use of the semi-arid zone resources cannot be achieved. The mere provision of food for the region's inhabitants, which has been for

Plate 19.   Tehran with Mt Demavand. A semi-arid environment with high, well-watered mountains, securing abundant groundwater (courtesy: the Director, Department of Geography, Hebrew University of Jerusalem).

thousands of years the aim of the arid zone peoples, who aspired to the 'land of milk and honey'—is no longer an answer for the development of any region.

These samples have taught us that under equal, or almost equal basic conditions—the natural challenges—human societies which are entirely different from one another, have been formed during the last century; societies of different occupations, standards of living and ways of life, in spite of the natural common denominator. If we examine the reason for the existing difference between them, and the widening gap between societies that live in the semi-arid zone, the answer seems to be the same as the one presented at the beginning of this book in the words of Jean Gottmann: the progress of science and technology gives to *a nation that is willing and knows how to organise*, an immense power to mould its environment, to draw the maximum benefit from the natural elements and to create sources for subsistence in the most diversified geographical frameworks. The central problem is, then, educational: to confer on a certain society the wish and the knowledge how to organise. Knowledge can be passed on from country to country, and it can be turned, with varying degrees of success, into a deep and fundamental expertise; the most difficult part in this subject is to confer the will, the initiative, the awareness that man is actually the creator of his environmental conditions. The warfare against ignorance, superstitions, submission to fate; open-mindedness to new ideas, the will not to stay behind, independence from fettering tradition, daring to try and taking risks—all these turn a backward society into a developed society. May we be optimistic and believe that this is the course that will be taken by mankind?

Appendix

Definitions and classifications of arid zones

Empirical definitions

These interpret botanical or agricultural findings in terms of interrelations between the principal climatic elements. Since precipitation is the most important meteorological element, and the one which is measured most, it was first accepted as the principal and initial criterion for the definition of aridity; the data for evaporation, which are more difficult to measure or to calculate, were only considered later. Instead, the botanical elements were taken to cover up ignorance concerning evapotranspiration, and as an expression of climatic conditions. Recently, a finer distinction between rain intensities is taking the place of information concerning crude amounts of rain.

Another meteorological element, which is taken into consideration as a determinant of a region's arid nature, is temperature; as temperatures rise, the evaporation increases as well, and with it grow the water needs of the vegetation; a knowledge of temperatures, then, helps to determine more accurately the borderline of aridity.

The classical coefficient, which tries to express the botanical world in a climatic formula, is that of Köppen (1931). On the basis of a comparison between the actual botanical conditions and the data on precipitation and temperature, Köppen determined the borderline of aridity as a certain ratio of temperatures to precipitation. His *aridity coefficient*—the borderline between the semi-humid zone and the semi-arid zone—in regions where there is no definite rainy season, was determined by the equation: $P \leqslant (T+7).2$, when P is precipitation in centimetres and T is the temperature in degrees Celsius. The borderline between the semi-arid and the arid was determined by a ratio which is twice smaller, that is: $P \leqslant T+7$. Therefore, if we take, for instance, a region with rains all the year round, and an average annual temperature of 18°C (64°F), then if its rain amounts are 500 mm (20 in) or less, it will be included in the semi-arid zone; if, with the same temperature conditions, the annual rain amounts are 250 mm (10 in), it will be included in the arid zone.

Nevertheless, Köppen understood very well the different value of the summer rains (the hot season), as compared to the winter rains, in respect of their efficiency for the biosphere; he established, therefore, the coefficient of $P \leqslant (T + 14).2$ for a region with summer rains, and the coefficient of $P \leqslant 2.T$ for a region with definite winter rains. Thus, returning to the example of a region with an average annual temperature of 18°C (64°F), if the case is that of summer rains (for instance, the southern borderland of the Sahara) the average annual amount of rain, necessary for including the region in the borderland between the semi-humid zone and the semi-arid zone, is 640 mm (25 in), whereas if it is a region of winter rains, 360 mm (14 in) is the marginal amount, as is the case in the northern Negev. In this changed form of the formula Köppen takes into account the efficiency of precipitation according to summer evaporation. As stated, this formula is only empiric and not the result of actual measurements of evaporation proportions.

There is no point in discussing all the many coefficients that have been tried to determine the aridity margin (they amount to several dozens). But three in particular deserve mention:

(*a*) The coefficient of DE MARTONNE (1927, 1935). It was based on the formula: coefficient of aridity $= \dfrac{P}{T + 10}$ when the coefficient is calculated for the whole year. The improvement made by De Martonne in 1935, was, that instead of being calculated for a whole year, aridity could be calculated

for any number of months or days: the coefficient of aridity $= \dfrac{n.P}{t + 10}$ when n is the number of rainy days, and t the average temperature value for these rainy days. A cofficient of values between 30 to 20 indicates a semi-arid climate, whereas a coefficient of values of 20 or less indicates an arid climate.

(*b*) EMBERGER, 1932, made a further improvement in the De Martonne coefficient, by taking into account the monthly fluctuations of temperature during the year: his aridity coefficient was $\dfrac{100,P}{(M^2 - m^2)}$, when M indicates the maximum monthly temperature and m the minimum monthly temperature.

(*c*) A different type of coefficient, that relates the climatic conditions to the possibility of the existence of vegetation, is that (*Arid Zone Res.*, 21, 1963) based on the subsistence conditions of the vegetation in the Mediterranean climate. It is particularly important for the marginal regions of the arid zone, yet the precision of a definition becomes problematic, not in the core of a defined phenomenon but on its margins; and if there are differences between the various coefficients, they are tested not in the centre of the region, but on its margins.

In the last classification the meteorological elements are temperatures

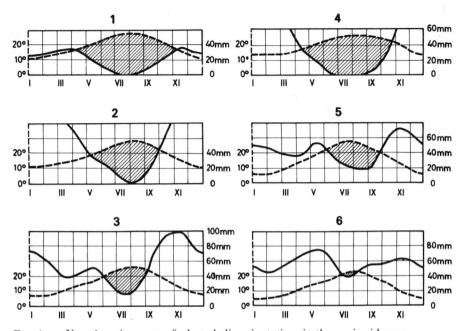

FIG. A1.   Xerothermic curves of selected climatic stations in the semi-arid zone
(Bioclimatic map of the Mediterranean Region, Unesco, 1963).
1. Marakesh, Morocco. 2. Seville, Spain. 3. Rome, Italy. 4. Beirut, Lebanon.
5. Larissa, Greece. 6. Toulouse, France. Rainfall – solid line; temperature – peaked line.
(100 mm = 3.9 in).

and the total amount of precipitation that the plant receives (from rain, snow, dew, fog and mist). The term 'arid month' refers to a month in which the total amount of precipitation (in millimetres) is less than, or equal to, twice the values of temperature of that same month in degrees Celsius. The classification is done by the Ombrothermic diagram. It presents temperature values on a double scale in relation to precipitation data (Fig. A1). The curve of precipitation values ('the Ombric curve') and the temperature curve ('the thermic curve') are drawn on the diagram; when the Ombric curve goes down below the thermic curve, it means that the amount of precipitation is smaller than twice the temperature values. The space between the curves represents proportionally the length and intensity of the dry season, and serves to determine the xerothermic coefficient (that is, the aridity coefficient and the temperature values attached to it). The xerothermic coefficient indicates the degree of aridity in a certain month, and is defined as *the number of days that may be regarded as arid from the biological point of view*. The classification derived is as follows:

The region is considered as having a desert climate, when the values of the xerothermic coefficient are above 300—that is, there are more than 300 days a year which are arid with regard to their biological value; it is an extreme desert climate when a whole year might pass without any rain at all, and the xerothermic coefficient is over 365; it is a subdesert climate when the coefficient values range between 100 and 300 (that is, the dry season is of nine to eleven months).

The arid regions with a dry season of one to eight months belong to the Mediterranean climate group (if the humid season is winter), and to the group of tropical climates (when the rains fall in the hot season). In the area of Mediterranean climates as defined above, there are four subregions:

1. *Xerothermomediterranean*, that is, hot and arid with a xerothermic co-efficient ranging between 200 and 150 (seven to five biologically arid months);
2. *Thermomediterranean*, when the coefficient is between 150 and 100 (that is, between five and three arid months).
3. *Mesomediterranean*, with a coefficient ranging between 100 and 40 (that is, 3 to 1½ arid months).
4. *Submediterranean*, when the coefficient is smaller than 40, and an absolute absence of a dry season is also possible; this type actually belongs to the humid side of the borderline.

Figure A1 illustrates these types of climates. Of all these climates the first three groups of the Mediterranean climate as defined above—a xero-thermic coefficient between 200 and 40—are the subject of this study.

Definitions based on the plant's water balance

Pioneer work in this field was done by Thornthwaite, in 1931, when the term 'effectiveness of precipitation' was coined. The method for determining the aridity of a region was developed by Thornthwaite in 1948 in the term 'potential evapotranspiration', that is, the total sum of evaporation and water consumption from an area which is densely covered by vegetation, provided there is a constant supply of available water. The elements included in the aridity coefficient—which is based on actual measurements of water consumption by plants— are precipitation, potential evaporation, water consumption (that is, evapotranspiration), water surplus, and water replenishment. Figure A2 illustrates these elements. From these thorough examinations Thornthwaite determined the aridity coefficient on the basis of the following formula:

$$\text{aridity coefficient} = \frac{\text{water surplus} . 100 - \text{water deficit} . 60}{\text{potential water needs}}$$

Different coefficient values present the borderlines between the semi-arid, the arid and the extreme arid climates.

At present Thornthwaite's is the most perfected classification of world climatic regions. It also serves as a basis for the detailed classification of arid zones, as done by Meigs in 1953, which is today the accepted classifica-

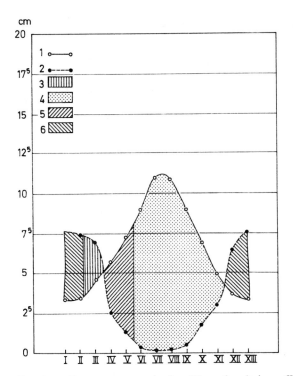

cm

FIG. A2. Parameters constituting Thornthwaite's coefficients of aridity, applied to Los Angeles (Thornthwaite, 1948). 1. Potential evapotranspiration. 2. Precipitation. 3. Water surplus. 4. Water deficiency. 5. Soil moisture utilisation. 6. Soil moisture recharge. On the ordinate, precipitation in centimetres; on the abscissa, months of the year.

tion of arid zones (Figs. A3–A8); the various arid regions have been defined in this work, by using this classification.

Meigs divides the arid zones into three types (which are fairly parallel to Thornthwaite's classification):

the *extreme arid climate*, where a total absence of rain for a whole year, or more, is possible;
the *arid climate*, approximately within the range of minus 20 to minus 40 according to Thornthwaite's classification;
the *semi-arid climate*, 0 to minus 20 according to Thornthwaite's classification.

In addition to this basic classification, Meigs classifies climatic regions according to their rainy season (summer, winter, or without a defined season), and also according to temperatures: mild, or extreme in summer, and the same for winter.

The possibilities of classification according to Meigs may, then, be illustrated in the following scheme:

ARIDITY	RAINY SEASON			TEMPERATURES (Celcius)							
	Winter c	Summer b	All year a	Winter				Summer			
				0–10	10–20	20–30	over 30	0–10	10–20	20–30	over 30
				1	2	3	4	1	2	3	4
EA: Extreme arid											
A: Arid	x			x							x
S: Semi-arid											

The classified region is designated by a capital letter (that indicates one of the three arid zones), a small letter, which represents the rainy season, and two figures—the first indicating values of winter temperatures and the second indicating summer temperature values (a high figure—extreme temperature). Thus, for instance, Ac13 (indicated by x in the table) shows an arid climate with winter rains, a rather cold winter and a hot summer. Such is, for instance, the climate of western Iran.

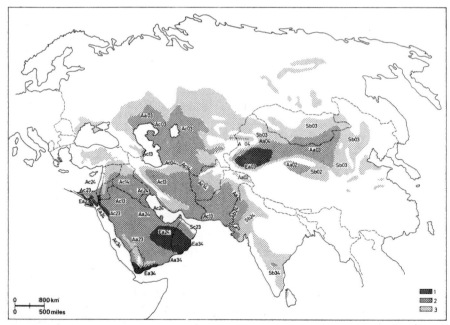

FIG. A3.   Arid climates in Asia. (McGinnies, 1968.) 1. Extreme arid. 2. Arid. 3. Semi-arid.

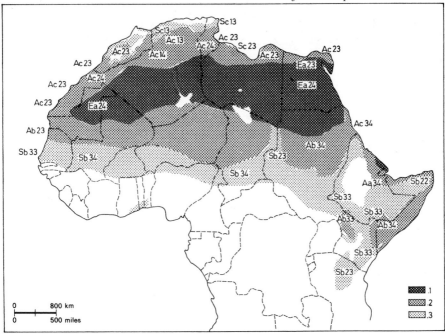

FIG. A4. Arid climates in North Africa. (McGinnies, 1968.) 1. Extreme-arid. 2. Arid. 3. Semi-arid.

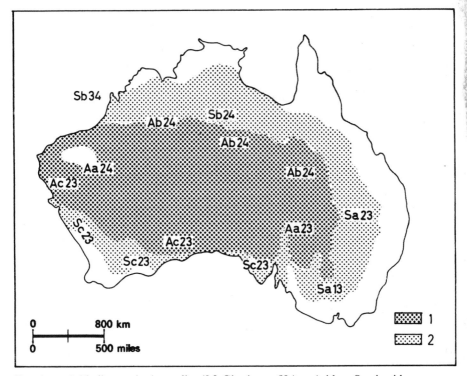

FIG. A5. Arid climates in Australia. (McGinnies, 1968.) 1. Arid. 2. Semi-arid.

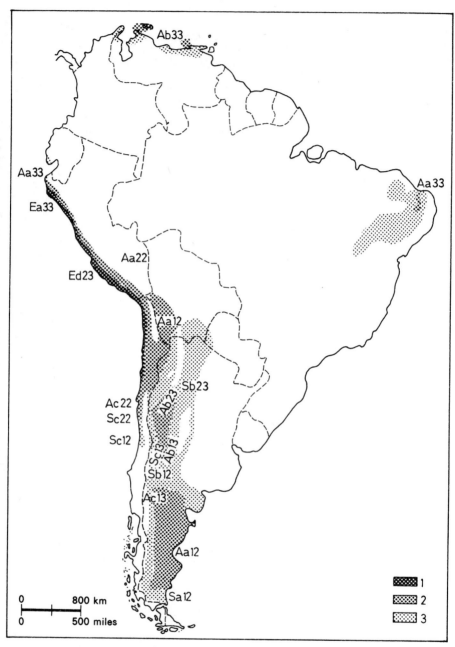

FIG. A6.   Arid climates in South America. (McGinnies, 1968.)
1. Extreme arid. 2. Arid. 3. Semi-arid.

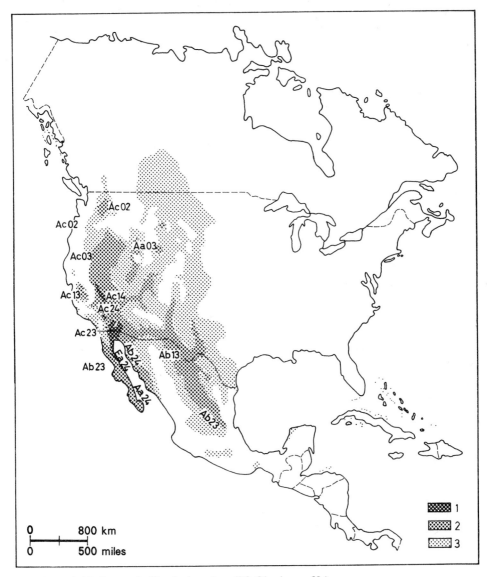

Fig. A7.   Arid climates in North America. (McGinnies, 1968.)
1. Extreme-arid. 2. Arid. 3. Semi-arid.

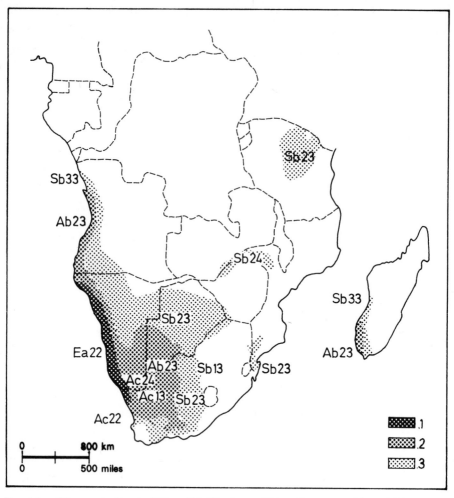

Fig. A8.   Climates in South Africa. (McGinnies, 1968.) 1. Extreme-arid. 2. Arid. 3. Semi-arid.

Geomorphological criteria for the classification of arid zones

The various aridity coefficients examine the arid zone from the angle of atmospheric conditions and their influence on the biosphere; yet there are also important criteria on the ground and the rocks that form the earth's crust. Since in the arid zone rocks are exposed with no continuous coverage of vegetation, the lithological variations of a certain area determine the processes of weathering and erosion, according to the contrasts between resistant and non-resistant rocks in arid conditions. It seems, therefore, that the geomorphological data enable an additional classification in the arid zone (Dresch, 1964; Tricart and Cailleux, 1969).

The principal geomorphological landscapes of the arid zone may be divided into two major groups: the planated arid zones of ancient shields, and a group of basins, piedmonts and the adjacent mountains.

The first group may include, for instance, the greater part of the Sahara, the Arabian peninsula, the Thar desert and the Australian desert. These are 'platform deserts'; the relief is very moderate, not usually reaching heights that permit the existence of less arid climates. Only a few mountains, such as the Hoggar and Tibesti, and the mountains of Sinai and Yemen, have semi-arid conditions, and experience frosts and snow. These platform deserts are identical with the ancient shields, that have developed in the course of hundreds of millions of years by processes of continental erosion. Their landforms are, generally, uniform: wide pediments, surmounted by residual mountains, or plateaus composed of sedimentary rocks, bounded by wide cuestas. A large part of these landscapes is a heritage from the humid Pleistocene, when lower temperatures and more active erosion deposited large accumulations of sand on the plains. Most of the landforms are erosional and not forms of accumulation; a great amount of crusts and coarse debris is found at the foot of escarpments.

The second group—that of basins, piedmonts and mountains, includes the Mediterranean regions, and the Alpine ranges, such as, the Andes, the Rocky Mountains, the basins of central Asia; these are basins dominated by the mountains that surround them. The mountains are generally sufficiently high for the formation of bioclimatic levels, which modify the arid conditions. The morphology of deserts, basins and piedmonts of this type, depends on the surrounding mountains and on the rivers and streams that originate in the melted snow that is available on these mountains, and whose power of transportation is sufficiently strong to carry the coarse debris which is formed on the mountain slopes. These basins are, then, mostly basins of deposition, and the erosional forms are scarce.

The distinction between the two geomorphological groups makes it possible to distinguish between the specific problems of the aforementioned regions; conditions for the existence of rivers, groundwater and soils are different for each group.

References

Key to abbreviations:

Ann. Ass. Am. Geogr.	Annals of the Association of American Geographers
Am. Ass. Adv. Sci.	American Association for the Advancement of Science
Annls. Géogr.	Annales de Géographie
Arid Zone Res.	Arid Zone Research (Unesco)
Bull. Ass. Géogr. Franc.,	Bulletin de l'Association Géographique de France
B.R.C.I.	Bulletin of the Research Council of Israel (Jerusalem)
C.O.M.	Cahiers d'Outre-Mer
Econ. Geogr.	Economic Geography
Geogr. Annlr.	Geografiska Annaler
Geogrl. J.	Geographical Journal
Geogrl. Rev.	Geographical Review
Geolog. Rundsch.	Geologische Rundschau
I.E.J.	Israel Exploration Journal
I.J.E.S.	Israel Journal of Earth Science
Inf. Géogr.	Informations Géographiques
J. Meteorol.	Journal of Meteorology
Méditer.	Méditerranée
R.G. Alpine	Revue de Géographie Alpine

AGNEW, S. (1959) 'South African farming and the pioneer legacy', in R. Miller, ed. *Geogr. Essays in memory of A. G. Ogilvie*, Nelson, pp. 221–46.

AHMAD, K. S. (1964) 'Land use in the semi-arid zone of Western Pakistan', *Arid Zone Res.*, **26**, 123–33.

AMIRAN, D. H. K. (1963) 'Effects of climatic change in an arid environment of land-use patterns', *Arid Zone Res.*, **20**, 437–42.

AMIRAN, D. H. K. (1964) 'Land use in Israel', *Arid Zone Res.*, **26**, 101–12.

AMIRAN, D. H. K. (1965) 'Arid zone development; a reappraisal under modern technological conditions', *Econ. Geogr.* **41**, 189–210.

AMIRAN, D. H. K. (1966) 'Man in arid lands', in Hills (1966) ed., pp. 219–54.

AMIRAN, D. H. K. and GILEAD, M. (1954) 'Early excessive rainfall and soil erosion in Israel', *Arid Lands in Perspective* (1969), University of Arizona Press, **4**, 286–95.

ASHBEL, D. (1950) *Bio-climate Atlas of Israel*, Meteorological Department of the Hebrew University, Jerusalem.

ASHGHAR, A. G. (1962) 'Public awareness and the educational problem in irrigation and land use in Pakistan', *Arid Zone Res.*, **18**, 429–75.

AUBERT, G. (1962) 'Arid zone soils', *Arid Zone Res.*, **18**, 115–37.

BATAILLON, C. ed. (1963) 'Nomads et nomadisme au Sahará, *Arid Zone Res.* **19**.

BATISSE, M. (1969) 'Problems facing arid-lands nations', in *Arid Lands in Perspective* (1969) pp. 3–12.

BEMONT, F. (1961) 'L'irrigation en Iran', *Annls Géogr.* **70**, 597–620.

BEN ARIEH, Y. (1965) *The Middle Jordan Valley*, Hakibbutz Hameuchad Publishing House, Tel Aviv, (Hebrew).

BHARADWAJ, O. P. (1961) 'The arid zone of India and Pakistan', *Arid Zone Res.* **17**, 143–74.

BIOCLAMATIC MAP OF THE MEDITERRANEAN ZONE, *Arid Zone Res.* **21**.

BODENHEIMER, F. S. (1953) 'Problems of animal ecology and physiology in deserts', in *Desert Research*, Research Council of Israel, Spec. Publ. no. 2, Jerusalem, 205–9.

BOMER, P. (1964) 'L'irrigation dans le bassin de l'Ebre (Espagne)', *Inf. Géogr.* **28**, 18–27.

BOWMAN, I. (1924) *Desert Trails of Atacama*, New York, A.G.S. Spec. Publ., no. 5.

BREMAUD, O. and PAGOT, J. (1962) 'Grazing lands, nomadism and transhumance in the Sahel', *Arid Zone Res.* **18**, 311–24.

BÜDEL, J. (1957) 'The Ice Age in the tropics', *Universitas* **1**, 183–91.

BUTZER, K. W. (1961) 'Climatic change in arid regions since the Pliocene', *Arid Zone Res.* **17**, 31–56.

BUTZER, K. W. (1963) 'The last pluvial phase of the eurafrican sub-tropics', *Arid Zone Res.* **20**, 211–21.

CANTOR, L. M. (1967) *A World Geography of Irrigation*, Oliver & Boyd.

CAPONERA, D. A. (1856) *Le droit des eaux dans les pays musulmans*, Rome, FAO.

CARRIÈRE, P. (1966) 'Le progrés de l'irrigation des terres dans la vallée moyenne du Syr-Daria (Plaine de Ferghana—Steppe de la Faime)', *Bull. Ass. Géogr. Franc.*, nos. 348–9, 15–33.

CHURCH, R. J. H. (1961) 'Problems and development of the dry zone of West Africa', *Geogrl.J.* **127**, 187–204.

COX, G. W. (1902) 'The artesian water-supply of Australia from a geographical stand point', *Geogrl.J.* **19**, 560–76.

CRESSEY, B. (1958) 'Quanats, karez and foggaras', *Geogrl.Rev.* **48**, 27–44.

DALMASSO, (1962) 'Une cité minière au Nord Sahara', Béchar-Djedid, Mediter., 22–49.

DAVEAU, S. (1968) *Vie pastorale, vie urbaine et vegetation en Mauritanie Sahélienne*, C. de Estudos Geográficos de Lisboa.

DEBENHAM, F. (1953) *Kalahari Sand*, Bell.

DEHMAN, M. and VERSTAPPEN, T. H. (1967) *Land Forms, Water and Land Use of the Quetta Area*, UNESCO Mission, Quetta (mimeograph).

DESPOIS, J. (1961) Development of land use in Northern Africa, *Arid Zone Res.* **17**, 219–237.

DESPOIS, J. (1964) 'Les paysages agraires traditionnels du Maghreb et du Sahara, septentrional', *Annls. Géogr.* **73**, 129–171.

DESPOIS, J. and RAYNAL, R. (1967) *Géographie de l'Afrique du nord-ouest*, Paris, Payot.

DIXEY, F. (1962) 'Geology and geomorphology, and groundwater hydrology', *Arid Zone Res.* **18**, 23–52.

DIXEY, F. (1966) 'Water supply, use and management', in Hills (1966), 77–102.

DOZIER, C. L. (1963) 'Mexico's transformed Northwest. The Yaqui, Mayo and Fuerte examples', *Geogrl. Rev.* **53**, 548–71.

DRESCH, J. (1956) 'Le Kyzyl-Koum et la sédentarisation des nomades', *Bull. Ass. Géogr. Franc.*, nos. 257–8, pp. 98–118.

DRESCH, J. (1964) 'Remarques sur une division géomorphologique des régions arides et les caractères originaux des régions arides mediterranéennes', *Arid Zone Res.* **26**, 23–30.

DUBIEF, J. (1963) Contribution au problème des changements de climats survenus aux cours de la période couverte par les observations météorologiques faites dans le Nord de l'Afrique', *Arid Zone Res.* **20**, 75–9.

EVENARI, M. and KOLLER, D. (1956) 'Desert agriculture: problems and results in Negev Desert of Israel', *Rehovot Nat. Agr. Inst.* **1**, (1963), 142 p., **2** (1964), 80 p.

EVENARI, M., SHANAN, L. and TADMOR, H. N. (1963–64) 'Runoff farming in the Negev Desert of Israel', *Rehovot* **1**, (1963), 142 p., **2** (1964), 80 p.

FLACH, K. W. and SMITH, G. D. (1969) 'New system of soil classification', in *Arid Lands in Perspective* (1969), pp. 59–73.

FLOHN, H. (1952): 'Allgemeine atmosphärische zirkulation und palao-klimatologie', *Geolog. Rundsch.* **40**, 153–78.

GARNSEY, M.E. and WOLLMAN, N. (1963) 'Economic development of arid regions', in Hodge (1963), pp. 369–95.

GEORGE, P. (1956) L'oasis de Tachkent, *Bull. Ass. Géogr. Franc.*, nos. 257–8, pp. 85–97.

GILAT, T. (1962) 'Diseases due to hot climate', Symposium *Climate and Man in Israel* held at the Negev Institute for arid zone research, Beer-Sheva, Israel, 4–5, National Council for research and development, Jerusalem, pp. 94–100. (Hebrew)

GOLDSCHMIDT, M. J. and JACOBS, M. (1958) 'Precipitation over and replenishment of the Yarqon and Nahal Hatteninim underground catchments', *Hydr. Paper* no. 3, Jerusalem, Hydrological Service.

GOLDING, E. W. (1962) 'Energy from wind and local fuels', *Arid Zone Res.* **18**, 249–58.

GOTTMANN, JEAN (1966) *Essais sur l'aménagement de l'espace habité,* Paris, Mouton.

GOUROU, P. (1966) 'Civilisation et désert, *L'Homme* **6**, 112–19.

GOUROU, P. (1970) *L'Afrique*, Paris, Hachette.

GRANDET, CL. (1958) 'La vie rurale dans le cercle de Goundam', V **11**, 25–46.

GRENIER, P. (1960) 'Les Peuls du Ferlo', *C.O.M.* **13**, 28–58.

GROVE, A. T. and WARREN, A. (1968) 'Quaternary landforms and climate on the south side of the Sahara', *Geogrl. J.* **134**, 194–208.

GUILCHER, A. (1965) *Précis d'hydrologie marine et continentale*, Paris, Masson, 389 p.

HENDERSON, D. A. (1965) 'Arid lands under agrarian reform in northwest Mexico' *Econ. Geog.* **41**, 300–12.

HILLEL, D. (1959) *Researches on Loess Crusts*, Agricultural Research Station, Beit Dagon, no. 63.

HILLS, E. S. ed. (1966) *Arid Lands*, Methuen/Unesco.

HODGE, C. ed. (1963) *Aridity and Man*, Am.Ass.Adv.Sci., Washington, D.C. 584 p.

HOWE, D. E. (1962) 'Saline water conversion', *Arid Zone Res.* **18**, 271–97.

HUMLUM, J. (1959) *La Géographie de l'Afghanistan*, Copenhagen.

ISSAR, A. and ECKSTEIN, Y. (1969) 'The lacustrine beds of Wadi Feiran, Sinai; their origin and significance', I.J.E.S. **18**, 21–7.

JACKSON, W. A. D. (1962) 'The virgin and the idle lands program reappraised', *Ann. Ass. Am. Geogr.* **52**, 69–79.

JAEGER, E. C. (1957) *The North-American Deserts*, Stanford University Press.

KARMON, Y. (1959) 'Geographic conditions in the Sharon Plain and their impact on its settlement', *Bulletin of the Israel Exploration Society* **23**, 3–24 (Hebrew).

KARSTEN, A. A. (1963) 'The Virgin Lands Kray and its prospects of development', *Soviet Geography*, May 1963, 37–46.

KEDAR, Y. (1967) *Ancient Agriculture in the Negev*, Bialik Institute, Jerusalem (Hebrew).

KÖPPEN, W. (1931) *Grundriss der Klimakunde*, Berlin.

KUBIENA, W. L. (1963) 'Paleosols as indicators of paleoclimates', *Arid Zone Res.* **20**, 207–8.

de LABURA, C. R. (1964) 'L'utilisation du sol dans la region semi-aride de l'Espagne', *Arid Zone Res.* **26**, 75–80.

LEE, H. K. D. (1962) Application of human and animal physiology and ecology to arid zone problem, *Arid Zone Res.* **18**, 213–33.

LEE, H. K. D. (1969) Variability in human response to arid environments, in *Arid Lands in Perspective*, pp. 229–45.

LEE, R. B. and VORE, I. DE, eds. (1968) *Man the Hunter*, University of Chicago Press.

LEOPOLD, L. B., WOLMAN, M. G. and MILLER, J. P. (1964) *Fluvial Processes in Geomorphology*, San Francisco, Freeman.

LEWIS, R. A. (1962) 'The irrigation potential of Soviet Central Asia', *Ann.Ass.Am. Geogr.* **52**, 99–114.

LOGAN, R. F. (1969) Geography of Central Namib desert, in *Arid Lands in Perspective*, pp. 129–44.

MARCUS, J. H. (1962) 'The problem of fatigue in a hot climate', Symposium, *Climate and man in Israel*, National Council for Research and Development, pp. 121–17 (Hebrew).

MARSHALL, L. (1960) 'Kung Bushman bands', *Africa* **30**, 325–55.

MARTONNE, E. DE (1927) 'Regions of interior-basin drainage', *Geogrl Rev.* **17**, 397–414.

McGINNIES, W. G., GOLDMAN, B. J. and PAYLORE, P., eds. (1968) *Deserts of the world: an appraisal of research into their physical and biological environments*, University of Arizona Press, Tucson.

McGINNIES, W. G. (1969) Arid lands knowledge gaps and research needs, in *Arid Lands in Perspective* (1969), pp. 279–87.

MEIGS, P. (1953) 'World distribution of arid and semi-arid homoclimates', *Arid Zone Res.* **1**, 203–9.

MEIGS, P. (1966) 'Geography of coastal deserts', *Arid Zone Res.* **28**, 140.

MEIGS, P. (1969) 'Future use of desert seacoasts', in *Arid Lands in Perspective* (1969), pp. 101–18.

MILTHORPE, F. L. (1960) 'The income and loss of water in arid and semi-arid zones', *Arid Zone Res.* **15**, 9–36.

MONGEIG, P. (1966) 'Les franges pionnières', Geogr. Gen., *Encyclopedie de la Pleiade*, Paris, pp. 976–1006.

NAYLON, J. (1967) 'Irrigation and internal colonization in Spain', *Geogrl J.* **133**, 178–91.

NAYLON, J. (1973) 'An appraisement of Spanish irrigation and land-settlement policies since 1939', *Iberian Studies* **2**, no. 1.

NELSON, H. J. (1959) 'The spread of an artificial landscape over Southern California', *Ann.Ass.Am.Geogr.* **49**, Suppl. pp. 80–100.

NIR, D. (1962) 'Trois ans de sécheresse consecutive dans la région de Beth-Chéane', *Mediter.* **3**, 3–19.

NIR, D. (1967) *Landscapes of Spain*, Massada, Ramat-Gan (Hebrew).

NIR, D. (1968) *La Vallée de Beth-Chéane—la mise en valeur d'une région a la lisière du désert*, École Pratique des Hautes Etudes, VI section, Etudes et Memoires, no. 65, Paris, Colin.

NIR, D. (1970) *Geomorphology of Israel*, Jerusalem, Academon (Hebrew).

OPPENHEIMER, H. R. (1960) 'Adaptation to drought; Xerophytism', *Arid Zone Res.* **15**, 105–38.

PAPY, L. (1959) 'Le déclin des foggaras au Sahara', *C.O.M.* **12**, 401–6.

PARDÉ, M. (1951) 'Sur le mecanisme des transports solides effectués par les rivières et sur les alterations correlatives des lits fluviaux', *R. G. Alpine* **39**, 5–40; 289–315.

PÉGUY, CH. P. (1961) *Précis de climatologie*, Paris, Masson.

PICARD, L. (1952) 'Outline on groundwater geology in arid regions', B.R.C.I. **2**, 358–71.

PLANHOL, X. and ROGNON, P. (1970) *Les Zones tropicales arides et subtropicales*, Paris, Colin.

POWNALL, E. (1967) *The Thirsty Land*, Methuen.

QUEZAL, P. (1963) 'De l'application de techniques palynologiques à un territoire désertique; paléoclimatologie du Quaternaire récent au Sahara, *Arid Zone Res.* **20**, 243–239.

RAUP, H. F. (1959) 'Transformation of Southern California to a cultivated land', *Ann.Ass.Am.Geogr.* **49**, Suppl., 58–79.

REIFENBERG, A. (1950) *The Struggle Between Sown Land and Desert*, Bialik Institute, Jerusalem.

RODIER, J. (1964) *Régimes hydrologiques de l'Afrique noire à l'ouest du Congo*, Paris, ORSTOM.

RUMNEY, G. R. (1968) *Climatology and the World's Climates*, New York, Macmillan.

SANTOS, M. (1963) 'Les difficultés de dévéloppement d'une partie de la zone seche de l'état de Bahia; la vallée moyenne du fleuve Paraguacu', *Annls Géogr.* **72**, 314–39.

SCHICK, A. P. (1969) 'The storm of 11 March, 1966, in the southern Negev and its geomorphic significance', *Studies in the Geography of Israel* **6**, 20–52 (Hebrew).

SHMUELI, A. (1970) *The settlement of the Bedouin of the Judaea Desert*, Tel Aviv, Gomeh (Hebrew).

STODDART, D. R. (1969) 'Climatic geomorphology, review and reassessment', *Progress in Geography* **1**, 159–222.

STRETA, E. J. and MOSINO, A. (1961) *Distribution de las zonas arides de la Republica Mexicana*, Ingenieria Hidraulica, Mexico DF, 8 p.

TAAFFE, R. N. (1962) 'Transportation and regional specialisation: the example of Soviet Central Asia', *Ann.Ass.Am.Geogr.* **52**, 80–98.

TABOR, H. (1962) 'Solar energy', *Arid Zone Res.* **18**, 259–70.

TALBOT, W. J. (1961) 'Land utilisation in the arid regions of southern Africa, Part I: South Africa', *Arid Zone Res.* **17**, 299–331.

THOMAS, G. W. and BOX, T. W. (1969) 'Social and ecological implications of water importation into arid lands', in *Arid Lands in Perspective* (1969), pp. 363–74.

THOMAS, W. L., ed. (1959) 'Man, time and space in southern California', *Ann.Ass. Am.Geogr.* **49**. Suppl.

THORNTHWAITE, C. W. (1948) 'An approach toward a rational classification of climate', *Geogrl Rev.* **38**, 55–94.

TIXERONT, J. (1963) 'Relations des fluctuations climatiques avec l'hydrologie, l'agriculture et l'activité humaine en Afrique du Nord', *Arid Zone Res.* **20**, 429–36.

TOTHILL, J. D. ed. (1948) *Agriculture in the Sudan*, Oxford University Press.

TRICART, J. and CAILLEUX, A. (1969) *Traité de géomorphologie, IV, Le modelé des régions séches*, Paris, SEDES, 472 p.

VAN DER POST, L. (1962) *The Lost World of the Kalahari*, Penguin Books.

WADHAM, S. (1961) 'The problem of arid Australia', in *History of Land-use in the Arid Regions*, UNESCO, Paris, *Arid Zone Res.*, **17**, 339–362.

WALLÉN, C. C. (1967) 'Aridity definitions and their applicability', *Geogr. Annlr.* **49**, 367–85.

WHITE, G. F. (ed.) (1956) *The Future of Arid Lands*, Am.Ass.Adv.Sci., Publ. no. 43, Washington D.C.

WHITE, G. F. (1960) *Science and the Future of Arid Lands*, Paris, Unesco.

WHYTE, R. O. (1963) 'Consequences de modifications du climat pour la végétation spontanée et l'agriculture', *Arid Zone Res.* **20**, 387–402.

WILLET, C. (1949) 'Long period fluctuations of the general circulation of the atmosphere, *J. Meteorol.*' **6**, 34–54.

WOLLMAN, N. (ed.) (1962) *The Value of Water in Alternative Uses*, University of New Mexico Press.

ZOHAR, E. (1962) 'Water and salt metabolism in hot climate', Symposium, *Climate and man in Israel*, National Council of Research and Development, pp. 80–90 (Hebrew).

Index